RECONSIDERING LOGICAL POSITIVISM

In this collection of essays, one of the preeminent philosophers of science writing today offers a reinterpretation of the enduring significance of logical positivism, the revolutionary philosophical movement centered around the Vienna Circle in the 1920s and 1930s. Michael Friedman argues that the logical positivists (especially Carnap, Reichenbach, and Schlick) were philosophical revolutionaries, not so much for presenting a new version of empiricism (as often is thought to be the case), but rather by offering a new conception of a priori knowledge and its role in empirical knowledge.

The positivists, under the influence of late nineteenth- and early twentieth-century developments in the foundations of geometry, logic, and mathematical physics, effected a profound transformation of the Kantian conception of synthetic a priori principles. The result is a relativized conception of a priori principles, which evolve with the progress of empirical science itself, but continue nevertheless to serve as a background framework for empirical principles properly so-called. These relativized a priori principles secure both the objective validity and the intersubjective communicability of natural science.

This collection will be mandatory reading for any philosopher or historian of science interested in the history of logical positivism in particular or the evolution of modern philosophy in general.

Michael Friedman is Ruth N. Halls Professor of Arts and Humanities, Professor of History and Philosophy of Science, and Professor of Philosophy at Indiana University. His previous publications include *Foundations of Space-Time Theories* (1983) and *Kant and the Exact Sciences* (1992).

RECONSIDERING LOGICAL POSITIVISM

MICHAEL FRIEDMAN

Indiana University

CAMBRIDGE UNIVERSITY PRESS
Cambridge, New York, Melbourne, Madrid, Cape Town, Singapore, São Paulo

Cambridge University Press
The Edinburgh Building, Cambridge CB2 8RU, UK

Published in the United States of America by Cambridge University Press, New York

www.cambridge.org
Information on this title: www.cambridge.org/9780521624497

First published 1999

A catalogue record for this publication is available from the British Library

Library of Congress Cataloguing in Publication data

Friedman, Michael, 1947–

Reconsidering logical positivism / Michael Friedman.

p. cm.

ISBN 0-521-62449-5

1. Science – Philosophy. 2. Logical positivism. I. Title.
Q175.F894 1999
501 – dc21 98-49422
CIP

ISBN 978-0-521-62449-7 hardback
ISBN 978-0-521-62476-3 paperback

Transferred to digital printing 2007

In memory of
J. Alberto Coffa (1935–1984),
quien abrió el sendero

CONTENTS

In der gesamten Geschichte der Mathematik gibt es wenig Ereignisse, die für die Gestaltung des Erkenntnisproblems und für seine Weiterentwicklung von so unmittelbar und tief einschneidender Bedeutung gewesen sind, wie die Entdeckung der verschiedenen Formen der Nicht-Euklidischen Geometrie.

Damit war ... nicht nur ein einzelnes mathematisches Problem gestellt; es war vielmehr die Frage nach der 'Wahrheit' der Mathematik, ja die Frage nach der Bedeutung der Wahrheit *überhaupt* in einem neuen Sinne gestellt. Nicht nur das Schicksal der Mathematik, sondern selbst das der Logik schien jetzt davon abzuhängen, in welcher Richtung die Lösung gesucht und gefunden wurde.

Ernst Cassirer
Das Erkenntnisproblem in der Philosophie und Wissenschaft der neuren Zeit
Vierter Band: *Von Hegels Tod bis zur Gegenwart (1832–1932)*

PREFACE

Logical positivism is one of the central strands in the fabric of twentieth-century thought. Originating in Austria and Germany in the 1920s, during the exuberant "modernism" of the Weimar period, it was intimately intertwined with some of the most important scientific developments of the new century: in particular, with the development and propagation of Einsteinian relativity theory and with the great debates on the foundations of mathematics that culminated in Gödel's celebrated incompleteness theorems. Indeed, Einstein was on close terms with several of the leading members of the logical positivist movement – with Moritz Schlick, the founder and guiding spirit of the Vienna Circle, and Hans Reichenbach, the leader of the Berlin Society for Empirical Philosophy – and Gödel was himself a part-time participant in the Vienna Circle. Logical positivism also was actively involved with the revolutionary sociocultural and political struggles of the period and, in particular, with the movement for a *neue Sachlichkeit* in both society and the arts typified by the Dessau Bauhaus. Otto Neurath, a second leading member of the Vienna Circle, was especially involved with these struggles: he represented a scientifically oriented version of Austro-Marxism and even served as minister of economics in the short-lived Bavarian Soviet Republic in 1919. The logical positivist movement thus was not only identified with Einsteinian physics and modern abstract mathematics, but also with socialism, internationalism, and "red Vienna."

It is by no means surprising, therefore, that the logical positivist movement was very actively engaged with the other vocal philosophical movements of the time as well – with neo-Kantianism, with Husserlian phenomenology, and even with the "existential-hermeneutical" variant of phenomenology then being initiated by Martin Heidegger. For all of these movements took it upon themselves to venture a radical reform of German philosophy in which it would renew and reinvigorate itself in a "scientific" spirit, much as the sciences themselves had recently done. Both Schlick and Rudolf Carnap (the third leading member of the Vienna Circle) were influenced by, and attempted to define themselves against, the neo-Kantianism of Ernst Cassirer (the leading contemporary representative of the so-called Marburg School) and the phenomenology of Edmund Husserl. Reichenbach, too, in his early work on relativity theory, articulated a new conception of the a priori closely analogous to the sophisticated neo-Kantian reading of Einstein's theory being developed by Cassirer at virtually the same time. And Carnap, in the early 1930s, engaged in a well-known polemical exchange with Heidegger on the topic of "metaphysical pseudosentences." Before the intellectual migration following the Nazi seizure of power in 1933, in other words, logical positivism was simply one new voice among others in the European philosophical landscape. It had its own distinctive revolutionary ambitions, to be sure, but it still spoke the language of, and defined itself in interaction with, the other novel trends within the German philosophical tradition.

After 1933 the leading logical positivists emigrated to the English-speaking world. Carnap and Reichenbach settled permanently in the United States, where they exerted a tremendous influence on the development of postwar "analytic" philosophy. (Schlick, who had visited at Stanford in 1929 as the first emissary of the movement to the new world, was murdered at the University of Vienna by a deranged student in 1936; Neurath died in exile in England in 1945.) Their influence, together with that of Carnap's protégé Carl Hempel (who earlier had studied with Reichenbach at Berlin), was decisive in establishing the new subdiscipline we now call "philosophy of science." Carnap's influence, in particular, also extended much further: to the widespread application of logical and mathematical methods to philosophical problems more generally, especially in semantics and the philosophy of language. Indeed, as is well known, the ideas of the logical positivists exerted a very substantial influence well beyond the boundaries of professional philosophy, particularly in psychology and the social sciences. It is not too much to say, therefore, that twentieth-century intellectual life would be simply unrecognizable without the deep and pervasive current of logical positivist thought.

Yet logical positivism's influence and reputation has now been eroded dramatically. It has been eclipsed, in the English-speaking world, by more pragmatic and naturalistic philosophical tendencies associated with the names of W. V. Quine and Thomas S. Kuhn – to such an extent, in fact, that logical positivism now often serves more as an intellectual scapegoat than as an honorable philosophical opponent. And in the German-speaking world, even today, "logical positivism" still popularly functions as a *Schimpfwort.* What is the explanation for this remarkable turn of events? How has one of the most central and influential trends in our twentieth-century intellectual landscape fallen into such ill repute? It is perfectly normal, of course, that once-dominant intellectual movements come to be underappreciated by their immediate successors, particularly when such successors, as is so often the case, depict themselves as directly reacting against their predecessors. But this natural and expected course of events seems quite inadequate to explain the full force, one might almost say the violence, of the reaction in this case.

The first point to notice is that the very intellectual migration that spread the influence of logical positivism so widely also resulted, in the end, in its dilution and eventual degeneration. In the European context of the 1920s, logical positivism arose and developed as a powerful revolutionary force, deeply intertwined with the other revolutionary trends (in the sciences, in the arts, in politics, and in society) that made up what we now know as Weimar culture. The logical positivists aimed at nothing less than a total refashioning of philosophy as a whole that would definitively end the fruitless, and endless, controversies of traditional metaphysics on behalf of a new "scientific" enterprise in which continuous and cooperative progress could be made solving fundamentally technical problems. And they took their inspiration and their models for such a radical disciplinary refashioning from the breathtaking revolutionary developments simultaneously taking place in mathematical physics and the foundations of mathematics. Although the positivists were, of course, also well aware that there were powerful opposing forces, particularly within German philosophy, working in a quite contrary direction, these developments in the sciences themselves still inspired them, in the words of Carnap's *Aufbau* (1928a), in "the faith that this [scientific-philosophical] orientation belongs to the future."

But these radical intellectual ambitions, which extended far beyond the boundaries of philosophy as a discipline, could not be transplanted easily onto American soil. The revolutionary context and rhetoric of radical philosophical transformation, especially in light of its explicitly Marxist overtones, had to be forgotten quickly as the erstwhile "scientific philosophers" from Central Europe were embraced by more down-to-earth and pragmatically minded thinkers such as Charles Morris, W. V. Quine, Nelson Goodman, and Ernest Nagel. In addition, the revolutions in the sciences from which the positivists had taken their philosophical inspiration had now run their course as well. Einsteinian relativity was a well-established part of mainstream physics; the "foundations crisis" in modern mathematics had reached a denouement, of sorts, with Gödel's incompleteness theorems. Finally, and perhaps most importantly, the positivists were no longer confronted with their philosophical competitors and rivals within the European philosophical tradition, the competing radical reform movements of Marburg neo-Kantianism, Husserlian phenomenology, and the "existential hermeneutics" of Heidegger. In the more comfortable postwar years following the global defeat of National Socialism, and in the more comfortable climate of American pragmatism and commonsense empiricism, the positivists lost much of their revolutionary fervor. No longer militantly crusading for a reform of philosophy as a whole, for a new type of scientific philosophy, they instead became respectable (and domesticated) practitioners of the new subdiscipline of philosophy of science. And, despite the impressive gains in clarity of some of the logical analyses thereby produced, this subdiscipline had reached a relatively unexciting period of stasis by the late 1950s and early 1960s.

We have not yet arrived at an explanation for the full force of the post-positivist reaction, however, and a second consequence of the emigration and postwar assimilation of logical positivism within the English-speaking world is of central importance in this regard. Extracted from the German intellectual tradition constituting their original philosophical context, the positivists now became identified with a rather simpleminded version of radical empiricism. Indeed, this process had already begun in earnest with the publication of A. J. Ayer's (1936) extraordinarily influential popularization of the logical positivist movement, *Language, Truth and Logic.* At the very center of positivist thought, according to this picture, is the notorious principle of verifiability, the principle that only propositions having direct implications for sensory experience are "cognitively meaningful." All other propositions, not only those of traditional metaphysics, but also those of ethics and religion, for example, are now declared to be devoid of such meaning. And, by the same token, a naively empiricist conception of natural science, now conceived as the paradigm of cognitively meaningful discourse, is the natural and inevitable complement to this view; because all of our theories in the natural sciences, no matter how complex and abstract they may appear, are now understood as elaborate devices for recording and systematizing our sensory experiences. Natural science as a whole thus is understood, in the end, as simply the continuous accumulation of more and more observable facts.

It is no wonder, then, that logical positivism, so characterized and identified, provoked an extreme, and one might even say violent, reaction, not only within professional philosophy but also within the humanities more generally. It is no wonder, in particular, that Thomas Kuhn's (1962) eloquent protest against naively empiricist conceptions of the growth of science in *The Structure of Scientific Revolutions* found such welcome resonance within both professional philosophy and the humanities more generally. Nevertheless, as scholarly investigations of the past fifteen or twenty years into the origins of logical empiricism have increasingly revealed, such a simpleminded radically empiricist picture of this movement is seriously distorted. Our understanding of logical positivism and its intellectual significance must be fundamentally revised when we reinsert the positivists into their original intellectual context, that of the revolutionary scientific developments, together with the equally revolutionary philosophical developments, of their time. As a result, our understanding of the significance of the rise and fall of logical positivism for our own time also must be fundamentally revised.

This volume contains my main contributions to this recent reinterpretive project. Characteristic of my approach is the idea that when we take due account of the scientific and philosophical context within which logical positivism developed, we see that their central philosophical innovation is not a new version of radical empiricism but rather a new conception of a priori knowledge and its role in empirical knowledge. The positivists, in my view, were led by late nineteenth- and early twentieth-century developments in the foundations of geometry, logic, and mathematical physics to venture a profound transformation of the Kantian conception of synthetic a priori principles: principles that are necessary, certain, and unrevisable but also applicable to the natural world given in our sensible experience. Revolutions in mathematics and mathematical physics have shown that no such principles are absolutely fixed for all time, so the positivists thought, but we still need characteristically nonempirical principles, which, despite their tendency to be revised in periods of deep conceptual revolution, are nonetheless constitutive of the framework for natural scientific investigation (of the rules of the game, as it were) at a given time. For it is to such principles, at any given time, that natural scientific investigation owes its objective validity and intersubjective communicability. The logical positivists' attempt to articulate this new conception of what we might call *relativized* a priori principles and to describe, in detail, how such principles make objective scientific experience first possible is, to my mind, the enduring philosophical legacy of that movement. For the underlying idea of a relativized a priori constitutively framing the empirical advances of natural science is still, in my opinion, of central philosophical significance. (In Kuhn's conception of the rules or "paradigms" definitive of a given episode of "normal science," for example, we find a very clear contemporary reflection of this idea.)

I here discuss the origins and development of the positivists' new conception of the a priori in the period, roughly, from 1915 to 1935. Part One concentrates on their earliest work, most of which was completed before the founding of the Vienna Circle in 1922, on the philosophical significance of the new discoveries in geometry and physics resulting in Einstein's formulation of the general theory of relativity in 1915. We see how the new conception of the a priori emerges from precisely these mathematical–physical developments, and in interaction, at the same time, with contemporaneous philosophical ideas arising within Marburg neo-Kantianism and Husserlian phenomenology. Part Two concentrates on Carnap's (1928a) *Der logische Aufbau der Welt,* which, along with Ludwig Wittgenstein's (1922) *Tractatus Logico-Philosophicus,* constituted the twin pillars of positivist philosophy during the heyday of the Vienna Circle (1925–31). We see that the *Aufbau,* which is standardly taken to be emblematic of the positivists' supposed naive empiricism, is instead centrally concerned with a new conception of the objectivity of empirical scientific knowledge having fundamentally Kantian roots. Part Three concentrates on Carnap's (1934c) *Logical Syntax of Language.* We see how the new conception of relativized a priori principles discussed in Part One is combined with Carnap's logically based conception of scientific objectivity discussed in Part Two. Working in intimate connection with the revolutionary foundational developments of the late 1920s and early 1930s, Carnap here produces the most mature and developed articulation of this new conception within the positivist movement.

This volume collects essays originally published over a period of fifteen years. In preparing them for publication here, I have, for the most part, made only very small revisions (including primarily new footnotes, cross references, and so on) aimed at introducing greater coherence and consistency into the whole. In several cases, however, I have changed my views substantially in the interim and have come to believe that some of my earlier formulations involved significant errors. In two of these cases, affecting Chapter 1 and Chapter 6, I have added new Postscripts explaining my updated views of the matter. In a third case, affecting Chapter 7, Chapter 9 explains my corrected views.

The Introduction originally appeared as "The Re-evaluation of Logical Positivism," *Journal of Philosophy* 88 (1991): 505–19. It is here reprinted with the permission of the editors.

Chapter 1 originally appeared as "Critical Notice: Moritz Schlick, Philosophical Papers," *Philosophy of Science* 50 (1983): 498–514. It is here reprinted with the permission of the University of Chicago Press.

Chapter 2 originally appeared as "Carnap and Weyl on the Foundations of Geometry and Relativity Theory," *Erkenntnis* 42 (1995): 247–60. It is here reprinted with kind permission from Kluwer Academic Publishers.

Chapter 3 originally appeared as "Geometry, Convention, and the Relativized A Priori," in W. Salmon and G. Wolters, eds., *Logic, Language, and the Structure of Scientific Theories* (Pittsburgh: University of Pittsburgh Press, 1994), pp. 21–34. It is here reprinted with the permission of the University of Pittsburgh Press.

Chapter 4 originally appeared as "Poincaré's Conventionalism and the Logical Positivists," *Foundations of Science* 1 (1995–96), pp. 299–314. It is here reprinted with the permission of the editors.

Chapter 5 originally appeared as "Carnap's *Aufbau* Reconsidered," *Noûs* 21 (1987): 521–45. It is here reprinted with the permission of the editors.

Chapter 6 originally appeared as "Epistemology in the *Aufbau*," *Synthese* 93 (1992): 15–57. It is here reprinted with kind permission from Kluwer Academic Publishers.

Chapter 7 originally appeared as "Logical Truth and Analyticity in Carnap's 'Logical Syntax of Language,'" in W. Aspray and P. Kitcher, eds., *History and Philosophy of Modern Mathematics* (Minneapolis: University of Minnesota Press, 1988), pp. 82–94. It is here reprinted with the permission of the University of Minnesota Press.

Chapter 8 originally appeared as "Carnap and Wittgenstein's *Tractatus*," in W. Tait, ed., *Early Analytic Philosophy* (La Salle, Ill.: Open Court Publishing Company, 1997), pp. 19–36. It is here reprinted with the permission of the Open Court Publishing Company.

Chapter 9 was originally written for inclusion in a volume honoring Burton Dreben, *Future Pasts*, ed. J. Floyd and S. Shieh, to appear. It appears here with the permission of Juliet Floyd and Sanford Shieh.

In addition, several pages in the Postscript to Chapter 1 originally appeared in "Helmholtz's *Zeichentheorie* and Schlick's *Allgemeine Erkenntnislehre*," *Philosophical Topics* 25 (1997): 19–50; they are here reprinted with the permission of the editor. Several pages in the Postscript of Chapter 6 originally appeared in "Overcoming Metaphysics: Carnap and Heidegger," in R. Giere and A. Richardson, eds., *Origins of Logical Empiricism* (Minneapolis:

University of Minnesota Press, 1996), pp. 45–79; they are here reprinted with the permission of the University of Minnesota Press.

I began work on the philosophical history of logical positivism while completing my dissertation on the relationship between relativity theory and the philosophy of science during the years 1970–2. This dissertation later appeared as Friedman (1983). I am indebted to my advisor, Clark Glymour, for first stimulating and encouraging my interest in this topic.

During the next few years, I became acquainted with Burton Dreben and Warren Goldfarb, who shared my interest and continued to stimulate and encourage it. Shortly thereafter Thomas Ricketts also began working on the topic – especially on the philosophy of Carnap – and the four of us engaged in sharply drawn debates on the interpretation and significance of Carnap's contributions. These debates, with Goldfarb and Ricketts in particular, have continued throughout the years and constitute the most important background to the work presented here. Chapter 9, in this context, simply reports on the latest stage of this debate from my own point of view.

In the early 1980s, I became acquainted with J. Alberto Coffa, who had been working for years on the history of logical positivism from a somewhat different perspective. The influence of Coffa's pioneering work pervades the present collection; and I deeply regret that his premature death in 1984 abruptly cut short our philosophical friendship.

Also in the 1980s, I began working with Alan Richardson in connection with his dissertation on Carnap's *Aufbau*, which ultimately appeared as Richardson (1998). This experience, it is fair to say, was important for both of us; our productive collaboration has continued throughout the years, and its results appear frequently in the present collection. During this same period, I became acquainted with Werner Sauer's neo-Kantian approach to Carnap, which served as a central source of inspiration for both Richardson and myself.

Around 1990 I made the acquaintance of a number of other philosophers approaching logical positivism from an historical point of view. Richard Creath's work on Carnap has been particularly important; and our philosophical interaction (often involving Goldfarb, Ricketts, and Richardson as well) has been especially fruitful. I should also give special mention to Thomas Oberdan, who completed a dissertation with me on Schlick that he had earlier begun with J. Alberto Coffa; Thomas Ryckman, whose penetrating contributions to the philosophical history of the theory of relativity have been fundamental; and Thomas Uebel, whose work on Otto Neurath has constantly served to put in perspective my own preoccupation with Carnap.

For additional valuable suggestions and advice on this project I am indebted to André Carus, William Demopoulos, Graciela De Pierris, Gottfried Gabriel, Silvana Gambardella, Ronald Giere, Anil Gupta, Rudolf Haller, Don Howard, Peter Hylton, Alison Laywine, Joia Lewis Turner, Isaac Levi, Ulrich Majer, Robert Nozick, W. V. Quine, Abner Shimony, Hans Sluga, Friedrich Stadler, Howard Stein, Richard Warner, and Mark Wilson. Finally, I would like to acknowledge Paul Pojman for his work in preparing the index.

INTRODUCTION

It is now well over half a century since the heyday of the philosophical movement known as logical positivism or logical empiricism. Depending on how one counts, it is now approaching half a century since the official demise of this movement.[1] Since that demise, it has been customary to view logical positivism as a kind of philosophical bogeyman whose faults and failings need to be enumerated (or, less commonly, investigated) before one's favored new approach to philosophy can properly begin. Such an attitude toward logical positivism and its demise has been widely prevalent, not only in the narrower community of philosophers of science (who characteristically have proceeded against the background of Kuhn's well-known critique), but also in the broader philosophical community as well. With our increasing historical distance from logical positivism, however, a more dispassionate attitude also has begun inevitably to emerge. No longer threatened or challenged by logical positivism as a live philosophical option, it is becoming increasingly possible to consider this movement as simply a part of the history of philosophy, which, as such, can be investigated impartially from an historical point of view. Indeed, we have seen in recent years a veritable flowering of historically oriented reappraisals of logical positivism.[2]

1 I assume that this event took place sometime between the publication of Quine's (1951) "Two Dogmas of Empiricism" and that of Kuhn's (1962) *The Structure of Scientific Revolutions.*

2 I have in mind, for example, work in the United States by A. Coffa, R. Creath, B. Dreben, P. Galison, R. Giere, W. Goldfarb, D. Howard, T. Oberdan, A. Richardson, T. Ricketts, T. Ryckman, J. Lewis Turner, and H. Stein; in Great Britain by N. Cartwright, J. Cat, S. Haack, A. Hamilton, J. Skorupski, and T. Uebel; and on the Continent by H.-J. Dahms, G. Gabriel, R. Haller, M. Heidelberger, K. Hentschel, A. Kamlah, U. Majer, V. Mayer, C. Moulines, P. Parrini, J. Proust, H. Rutte, W. Sauer, F. Stadler, and G. Wolters. In addition, the *Vienna*

In the course of these reappraisals, it has become clear – not at all surprisingly, of course – that the above-mentioned postpositivist reaction gave birth to a large number of seriously misleading ideas about the origins, motivations, and true philosophical aims of the positivist movement. (One can hardly expect philosophical critics, concerned largely with their own agendas rather than with historical fidelity, to generate anything other than stereotypes and misconceptions.) I will discuss what I take to be some of the most important of such misleading ideas in what follows, but I also hope to show – or at least to suggest – that achieving a better understanding of the background, development, and actual philosophical context of logical positivism is not *merely* of historical interest. For the fact remains that our present situation evolves directly – for better or for worse – from the rise and fall of positivism, and what I want to suggest is that we will never successfully move beyond our present philosophical situation until we attain a properly self-conscious appreciation of our own immediate historical background.

I

Perhaps the most misleading of the stereotypical characterizations views logical positivism as a version of philosophical "foundationalism." The positivists – so this story goes – were concerned above all to provide a philosophical justification of scientific knowledge from some privileged, Archimedean vantage point situated somehow outside of, above, or beyond the actual (historical) sciences themselves. More specifically, they followed the lead of the logicist reduction of mathematics to logic, where the latter also is understood as fundamentally foundationalist in motivation and import. Just as the logicists attempted to justify mathematical knowledge and place it on a secure foundation by means of a derivation from (supposedly more certain) logical knowledge, so the positivists attempted to justify empirical science and place it on a secure foundation by logically constructing the concepts of empirical science on the basis of the (supposedly more certain) immediate data of sense. Thus, formal logic furnished the foundational enterprise with the required Archimedean standpoint located outside of the actual (historical) sciences themselves, and phenomenalist reductionism, carried out rigorously using the tools of formal logic (as epitomized

Circle Collection, under the general editorship of R. Cohen, recently has made available in English many previously untranslated works of Schlick, Neurath, Reichenbach, Menger, and Waismann. Unfortunately, Carnap's earlier works are not yet available in English translation.

in Carnap's [1928a] *Der logische Aufbau der Welt*), then provided the desired epistemological justification of the sciences.[3]

This conception of the aims and posture of philosophy vis à vis the special sciences represents an almost total perversion of the actual attitude of the logical positivists, who rather considered their intellectual starting point to be a *rejection* of all such philosophical pretensions. An eloquent example is found in the first paragraph of a 1915 paper by Schlick on the need for philosophy to adapt itself to the new findings of relativity theory:

> We have known since the days of Kant that the only fruitful method of all theoretical philosophy consists in critical inquiry into the ultimate principles of the special sciences. Every change in these ultimate axioms, every emergence of a new fundamental principle, must therefore set philosophical activity in motion, and has naturally done so even before Kant. The most brilliant example is doubtless the birth of modern philosophy from the scientific discoveries of the Renaissance. And the Kantian Critical Philosophy may itself be regarded as a product of the Newtonian doctrine of nature. It is primarily, or even exclusively, the principles of the exact sciences that are of major philosophical importance, for the simple reason that in these disciplines alone do we find foundations so firm and sharply defined, that a change in them produces a notable upheaval, which can then also acquire an influence on our world-view. (1915/1978–9, p. 153)

Schlick goes on to argue that neither of the two prevailing philosophical systems – neither the "neo-Kantianism" of Cassirer and Natorp nor the "positivism" of Petzoldt and Mach – can do justice to Einstein's new theory and *therefore both systems must be abandoned.* Entirely new philosophical principles, based on the work of Einstein himself and of Poincaré, are necessarily required.

For Schlick, then, philosophy as a discipline is in no way foundational with respect to the special sciences. On the contrary, it is the special sciences that are foundational for philosophy. The special sciences – more specifically, the "exact sciences" – simply are taken for granted as paradigmatic of knowledge and certainty. Far from being in a position somehow to justify these sciences

3 The most explicit example of this kind of characterization of logical positivism of which I am aware is found in Giere (1988, pp. 22–8). This kind of conception is certainly aided and abetted (although with important reservations in each case) by Quine's (1969) "Epistemology Naturalized" and Ayer's (1936) *Language, Truth and Logic.* Rorty has portrayed the entire modern philosophical tradition as based on "foundationalist epistemology," under which rubric the logical positivists also are clearly thought to fall (1979, pp. 59, 332–3), and it is thus clear for Rorty that Kuhn's critique is predicated on a rejection of such foundationalist epistemology.

from some higher vantage point, it is rather philosophy itself that is inevitably in question. Philosophy, that is, must *follow* the evolution of the special sciences so as to test itself and, if need be, to reorient itself with respect to the far more certain and secure results of these sciences. In particular, then, the central problem of philosophy is not to provide an epistemological foundation for the special sciences (they already have all the foundation they need), but rather to redefine its own task in the light of the recent revolutionary scientific advances that have made all previous philosophies untenable.[4]

Moreover, this conception of the proper stance of philosophy vis à vis the special sciences is not peculiar to Schlick; it is in fact characteristic of the logical positivists generally. This is so even – and indeed especially – of Carnap's *Aufbau* (which work, of course, is supposed to be most representative of foundationalist epistemology).[5] It is true that the *Aufbau* presents a phenomenalist reduction of all concepts of science to the immediately given data of experience. Yet the point of this construction has little if anything to do with traditional foundationalism. First, Carnap shows no interest whatever in the philosophical skepticism that motivates, for example, Russell (1914) in *Our Knowledge of the External World*, nor does the text of the *Aufbau* at any point engage the traditional vocabulary of "certainty," "doubt," "justification," and so on.[6] Second, and more important, Carnap is perfectly explicit that the particular constructions he employs depend entirely on the actual "results of the empirical sciences" (§122) and, accordingly, that the particular constructional system that he presents is best viewed as a "rational reconstruction" of the actual (empirical) process of cognition (§100).[7] Thus, for example, Carnap's choice of holistic "elementary experiences"

4 As Schlick indicates, this is also a fruitful way of viewing Kant's philosophical stance with respect to the special sciences. For Kant, too, it is philosophy itself rather than the special sciences that is in question and therefore requires justification. Kant's aim is not to ground the sciences in something firmer and more secure (for it is they that are paradigmatic of certainty) but rather to reform *metaphysics* in accordance with the already achieved successes of the exact sciences. See the Preface to the second edition of the *Critique of Pure Reason* (especially Bxxii–xxiii), §VI of the Introduction to the *Critique*, and §40 of the *Prolegomena*. This undercuts decisively, it seems to me, Rorty's (1979) portrayal of Kant as the *echt*-foundationalist.

5 Carnap (1928a/1967); references are given in the text by section numbers.

6 Carnap does employ this language in retrospectively describing the motivation of the *Aufbau* at several points in his "Intellectual Autobiography" (1963a, pp. 51, 57). Nevertheless, the contrast between this retrospective account and the language of the *Aufbau* itself is striking indeed. These passages from Carnap's "Intellectual Autobiography" are discussed in detail in Chapter 6 (this volume).

7 Compare Carnap (1963a, p. 18).

rather than atomistic sensations as the basis of his system is grounded in the empirical findings of Gestalt psychology (§67), his definition of the visual sense modality depends on the (supposed) empirical fact that it is the unique sense modality having exactly five dimensions (§86), and so on. In particular, then, the findings of the special sciences, such as empirical psychology, are in no way in question in the *Aufbau*: once again, it is philosophy that must adapt itself to them rather than the other way around.

The aim of the *Aufbau*, therefore, is not to use logic together with sense data to provide empirical knowledge with an otherwise missing epistemological foundation or justification. Its aim, rather, is to use recent advances in the science of logic (in this case, the Russellian type-theory of *Principia Mathematica*) together with advances in the empirical sciences (Gestalt psychology in particular) to fashion a scientifically respectable *replacement* for traditional epistemology. Carnap's depiction of the construction of scientific knowledge from elementary experiences via the logical techniques of *Principia Mathematica* enables us to avoid the metaphysical excesses of the traditional epistemological schools – "realism," "idealism," "phenomenalism," "transcendental idealism"(§177) – while simultaneously capturing what is correct in all of these schools: allowing us to represent, in Carnap's words, the "neutral basis [*neutrale Fundament*]" common to all (§178).[8]

II

According to the standard picture of logical positivism briefly sketched earlier, that movement is to be understood not only as a species of foundationalist epistemology but also as a version of *empiricist* epistemology in the tradition of Locke, Berkeley, Hume, Mach, and Russell's external-world program.[9] The immediately given data of sense are viewed as the primary examplars of knowledge and certainty, and all other putative claims to knowledge are judged to be warranted – or, as the case may be, unwarranted – in light of their relations to the immediate data of sense. (When empiricist reductionism is acknowledged to have failed, the positivists' naive empiricism then is thought to express itself in the doctrine of an epistemically privileged and theory-neutral observation language against which all scientific claims are to be tested.) I dealt earlier with the conception of the positivists as

8 For details of this kind of approach to the *Aufbau*, see Part Two (this volume) and Richardson (1990, 1992, 1998).

9 This is, of course, the conception promulgated especially in Ayer's (1936) *Language, Truth and Logic*: compare the first sentence of Ayer's Preface, where the views put forward are characterized as "the logical outcome of the empiricism of Berkeley and David Hume."

foundationalists, and this question is, I think, rather easily disposed of. The question of empiricism – and, in particular, of the relationship between the positivists and the traditional empiricism of Locke, Berkeley, Hume, and Mach – is, however, considerably more delicate.

The first point to notice is that the positivists' main philosophical concerns did not arise within the context of the empiricist philosophical tradition at all. Rather, the initial impetus for their philosophizing came from late nineteenth-century work on the foundations of geometry by Riemann, Helmholtz, Lie, Klein, and Hilbert – work that, for the early positivists, achieved its culmination in Einstein's theory of relativity.[10] The principal philosophical moral that the early positivists drew from these geometrical developments was that the Kantian conception of pure intuition and the synthetic a priori could no longer be consistently maintained – consistently, that is, with the situation now presented to us by the exact sciences. In particular, Hilbert's (1899) logically rigorous axiomatization of Euclidean geometry shows conclusively that spatial intuition has no role to play in the reasoning and inferences of pure geometry, and the development of non-Euclidean geometries together with their actual application to nature by Einstein show conclusively that our knowledge of geometry cannot be synthetic a priori in Kant's sense. All the early positivists were thus in agreement that the strictly Kantian conception of the a priori must be rejected, and this rejection of the synthetic a priori constituted a centrally important element in what they came to call their "empiricism."

Yet it is equally important to notice, in the second place, that the positivists did not react to the demise of the Kantian synthetic a priori by adopting a straightforwardly empiricist conception of physical geometry of the kind traditionally imputed to Gauss (who is reported to have attempted to determine the curvature of physical space by measuring the angle sum of a terrestrial triangle determined by three mountaintops). On the contrary, all the early positivists *also* strongly rejected this kind of empiricist conception (which they attributed to Gauss, Riemann, and, at times, Helmholtz) and, rather, followed the example of Poincaré in maintaining that there is no direct route from sense experience to physical geometry: essentially nonempirical factors, variously termed "conventions" or "coordinating definitions," must necessarily intervene between sensible experience and geometrical theory.

10 Most of the early writings of the positivists focused on these revolutionary mathematical–physical developments. In addition to the 1915 paper of Schlick cited earlier, see Schlick (1917/1978), Reichenbach (1920/1965), and Carnap (1922). Interest in these themes also was stimulated by Weyl's 1919 edition of Riemann's 1854 dissertation and the 1921 edition of Helmholtz's epistemological writings by Schlick and the physicist Paul Hertz.

The upshot is that it is in no way a straightforward empirical matter of fact whether space is Euclidean or non-Euclidean.[11]

This radically new conception of physical geometry – neither strictly Kantian nor strictly empiricist – was formulated by Reichenbach (1920) in an especially striking fashion in his first book, *The Theory of Relativity and A Priori Knowledge.* Reichenbach maintains a sharp distinction, within the context of any given scientific theory, between two intrinsically different types of principles: "axioms of connection" are empirical laws in the traditional sense recording inductive regularities involving terms and concepts that are already sufficiently well defined; "axioms of coordination," on the other hand, are nonempirical statements that must be antecedently laid down before the relevant terms and concepts have a well-defined subject matter in the first place. (Thus, for example, Gauss's attempt empirically to determine the curvature of space via a terrestrial triangle inevitably fails because it tacitly presupposes that light rays travel in straight lines and, therefore, that the notion of a "straight line" is already well defined. But how, independently of the geometrical and optical principles supposedly being tested, can this possibly be done?) These nonempirical axioms of coordination – which include, paradigmatically, the principles of physical geometry – are thus "constitutive of the object of knowledge," and, in this way, we can therefore vindicate *part* of Kant's conception of the a priori.

For, according to Reichenbach, the a priori had two independent aspects in Kant: the first involves necessary and unrevisable validity, but the second involves only the just-mentioned feature of "constitutivity." The lesson of modern geometry and relativity theory, then, is not that the Kantian a priori must be abandoned completely, but rather that the constitutive aspect must be separated from the aspect of necessary validity. Physical geometry is indeed nonempirical and constitutive – it is not itself subject to straightforward observational confirmation and disconfirmation but rather first makes possible the confirmation and disconfirmation of properly empirical laws (viz., the axioms of connection). Nevertheless, physical geometry can still evolve and change in the transition from one theoretical framework to another: Euclidean geometry, for example, is a priori in this constitutive sense in the context of Newtonian physics, but only topology (sufficient to *admit* a Riemannian structure) is a priori in the context of general relativity.

11 See especially Poincare's (1902/1905, p. 79) *La Science et l'Hypothèse*: "Whichever way we look at it, it is impossible to discover in geometrical empiricism a rational meaning." Einstein himself frequently expressed admiration for Poincaré's doctrine: see, e.g., Einstein (1921/1923).

Reichenbach concludes that traditional empiricism is in error in not recognizing the a priori constitutive role of axioms of coordination, and it is clear, moreover, that, despite some terminological wrangling on this point, the other logical positivists are in substantial agreement.[12] One of the central themes of Schlick's (1918) *General Theory of Knowledge*, for example, is a sharp dichotomy between raw sensible acquaintance and genuine objective knowledge. Immediate contact with the given is both fleeting and irredeemably subjective. Objective knowledge therefore requires *concepts* and *judgments*, which are to be carefully distinguished from intuitive sensory presentations. Concepts and judgments are, in fact, only possible in the context of a rigorous formal system, of which Hilbert's axiomatization of geometry is paradigmatic. More specifically, the Hilbertian notion of "implicit definition" of scientific concepts via their logical places in a formal system (which notion is now associated by Schlick with Poincaré's conventionalist philosophy of geometry) can alone explain how rigorous, exact, and truly objective representation is possible. Here, we are obviously very far from traditional empiricism and very close indeed to the supposedly antipositivist doctrine of the theory ladenness of observation.[13]

Once again, the same is true even – indeed, especially – of Carnap's *Aufbau*. Carnap outlines an elaborate construction of all scientific concepts from the immediately given data of sense using the logical machinery of *Principia Mathematica*. Yet he also holds that the objective meaning of scientific concepts can in no way depend on merely ostensive contact with the given. On the contrary, intersubjective communication is possible only in virtue of the *logical structure* of the concepts in question arising from their logical places within the total system of scientific knowledge. More specifically, intersubjective meaning must derive entirely from what Carnap calls "purely structural definite descriptions" rather than from sensory ostension (§§12–15). (Thus, for example, the visual sense modality is defined in terms of the purely formal properties of its dimensionality rather than in terms of its phenomenal content.) In this sense, the elaborate logical structure erected

12 The terminological wrangling in question arises in correspondence between Schlick and Reichenbach in November 1920 concerning Reichenbach's book. Schlick criticizes Reichenbach for conceding too much to the Kantian side and argues that Reichenbach's a priori constitutive principles should rather be understood as conventions *à la* Poincaré. This correspondence is discussed in Chapter 3 (this volume). Schlick himself alludes to it (1921/1978, p. 333).

13 For further discussion of Schlick, see Chapter 1 (this volume). Theory ladenness follows from the circumstance that *all* scientific concepts, including those used to report the results of experiments, acquire objective meaning from their places in a theoretical system. For further discussion of this point, see Friedman (1992c).

above the basic elements of the system (the elementary experiences) is actually more important than the basic elements themselves. Objective meaning flows from the top down, as it were, rather than from the bottom up.[14]

Carnap illuminatingly articulates the precise meaning of the resulting "empiricism" in the Preface to the second edition (1961) of the *Aufbau*:

> For a long time, philosophers of various persuasions have held the view that all concepts and judgments result from the cooperation of experience and reason. Basically, empiricists and rationalists agree in this view, even though both sides give a different estimation of the two factors, and obscure the essential agreement by carrying their viewpoints to extremes. The thesis which they have in common is frequently stated in the following simplified version: The senses provide the material of cognition, reason works up [*verarbeitet*] the material so as to produce an organized system of knowledge. There arises then the problem of finding a synthesis of traditional empiricism and traditional rationalism. Traditional empiricism rightly emphasized the contribution of the senses, but did not realize the importance of logical and mathematical forms.... I had realized, on the one hand, the fundamental importance of mathematics for the formation of a system of knowledge and, on the other hand, its purely logical, purely formal character to which it owes its independence from the contingencies of the real world. These insights formed the basis of my book.... This orientation is sometimes called "logical empiricism" (or "logical positivism"), in order to indicate the two components. (1928a/1967, pp. v–vi)

In thus emphasizing the central importance of a priori formal elements in first providing objective meaning for the otherwise undigested immediate data of sense, logical positivism has, it seems to me, broken decisively with the traditional empiricism of Locke, Berkeley, Hume, and Mach – which tradition is understood by the positivists themselves as mistakenly giving epistemic centrality to precisely such undigested immediate sensory data. Perhaps the best way to put the point is that the logical positivists have staked out an entirely novel position that is, as it were, intermediate between traditional Kantianism and traditional empiricism: it gives explicit recognition to the constitutive role of a priori principles, yet, at the same time, it also rejects the Kantian characterization of these principles as *synthetic* a priori.

III

The novel philosophical position briefly sketched above, however, is faced with several fundamental problems. The heart of the positivists' new

14 For further discussion, see the references cited in footnote 8, this chapter.

conception is the idea of constitutive but nonsynthetic a priori principles underlying the possibility of genuinely objective scientific knowledge. There is a sharp distinction, in particular, between conventions, coordinating definitions, or axioms of coordination, on the one hand, and empirical principles properly so-called, on the other. On the one side lie the principles of pure mathematics and, at least in the context of some physical theories, the principles of physical geometry; on the other side lie standard empirical laws such as Maxwell's equations and the law of gravitation. But what is the basis for this distinction, and how, more generally, are we sharply to differentiate the two classes of principles? On a strictly Kantian view, such a distinction is, of course, grounded in the fixed constitution of our cognitive faculties, and the question of differentiation is correspondingly straightforward: the a priori principles are precisely those possessing necessary and unrevisable validity.[15] Now, however, we have explicitly acknowledged that a priori constitutive principles possess no such necessary validity and, in fact, that these principles may evolve and change – in response to empirical findings – with the progress of empirical science. So what exactly distinguishes our a priori principles from ordinary empirical laws properly so-called?

A second problem is perhaps even more fundamental. The logical positivists, I have argued, strongly rejected a foundationalist conception of philosophy vis à vis the special sciences. There is no privileged vantage point from which philosophy can pass epistemic judgment on the special sciences: philosophy is conceived rather as following the special sciences so as to reorient itself in response to their established results. But what then *is* the peculiar task of philosophy, and how, in general, does it relate to the special sciences? Is philosophy itself simply one special science among others, and, if not, from what perspective does it then respond to and rationally reconstruct the results of the special sciences? The positivists are nearly unanimous in explicitly rejecting a naturalistic conception of philosophy as simply one empirical science among others – a branch of psychology, perhaps, or of the sociology of knowledge.[16] On the whole, they instead prefer

15 Compare Kant's well-known "criterion by which to distinguish with certainty between pure and empirical knowledge" articulated in §II of the Introduction to the *Critique of Pure Reason.*

16 The conspicuous exception here is Neurath, who articulates a version of naturalism bearing some similarities to well-known views of Quine. Neurath, unlike the other members of the Vienna Circle, approaches philosophical questions from a background in the social sciences rather than in the mathematical exact sciences, and, accordingly, he shows very little interest in either the problem of a priori principles or the problem of elucidating the peculiar position and role of philosophy vis à vis the special sciences. See Haller (1993), Uebel (1992, 1996), and the essays in Uebel (1991).

to think of philosophy as in some sense a branch of *logic* and to conceive the peculiarly philosophical task as that of "logical analysis" of the special sciences. Yet the perspective or point of view from which such logical analysis is to proceed remains radically unclear.

No real answer to these questions was forthcoming until Carnap's (1934c) *Logical Syntax of Language*. Here, Carnap is once again responding to recent developments in the exact sciences: to Heyting's formalization of intuitionistic arithmetic, for example, and, above all, to Hilbert's program of "metamathematics." Moreover, he is once again attempting to neutralize the philosophical disputes arising in connection with these developments by showing how all parties involved are in possession of *part* of the truth; the remaining part that appears to be in dispute then is argued to be not subject to rational debate at all. More specifically, the dispute is declared to be a matter of convention in precisely Poincaré's sense: there is simply no fact of the matter concerning which party is "correct," and thus the choice between them is merely pragmatic.[17]

The dispute in question arises from increasing appreciation of how fundamentally the program of *Principia Mathematica* is threatened by the paradoxes and involves the three traditional schools in the foundations of mathematics: logicism, formalism, and intuitionism. Carnap responds to this dispute by declaring that each side is simply putting forth a proposal to construct a formal system or calculus of a certain kind: logicism proposes to construct an axiomatization of mathematics using the rules of classical logic, intuitionism proposes to construct an axiomatization using the more restrictive rules of intuitionistic logic, and so on. The essential point is that no such formal system or calculus is more correct than any other – indeed, the notion of correctness is entirely inappropriate here. Considered simply as proposals to construct formal systems of various kinds, all of the apparently opposing philosophies are then equally correct, and the choice between them can only be a purely pragmatic question of convenience. In the Foreword to *Logical Syntax*, Carnap calls this standpoint the *Principle of Tolerance* and then remarks:

> The first attempts to cast the ship of logic off from the *terra firma* of the classical forms were certainly bold ones, considered from the historical point of view. But they were hampered by the striving after "correctness." Now, however, that

17 Compare again Poincaré (1902/1905, p. 50): "What, then, are we to think of the question: Is Euclidean geometry true? It has no meaning. . . . One geometry cannot be more true than another; it can only be more convenient."

> impediment has been overcome, and before us lies the boundless ocean of unlimited possibilities. (1934c/1937, p. xv)

It is thus with an exuberant sense of liberation that Carnap extends Poincaré's conventionalism to logic itself.

But how is it possible for Carnap to maintain a stance of neutrality with respect to logic itself? Here is where the fundamental insights of Hilbert's metamathematics come into play. For we are to describe the logical rules governing the formal systems or calculi under consideration within the *metadiscipline* of logical syntax: each system is viewed simply as a set of strings of symbols together with rules for manipulating such strings (entirely independently of any question concerning their "meanings"), and any and all fomal systems then can be specified from the neutral standpoint of a purely syntactic meta-language. More precisely, the syntactic meta-language need employ only the limited resources of primitive recursive arithmetic, and we thus can describe the rules of classical systems, intuitionist systems, and so on from a standpoint that is neutral between them. Our aim is not to justify one system over others as inherently more correct, but simply to describe the *consequences* of choosing any such system.

Carnap is now in a position precisely to articulate the method and stance of logical analysis. This paradigmatically philosophical enterprise is simply a branch of logical syntax: specifically, the logical syntax of the language of science.[18] We thus are concerned with what Carnap calls "the physical language" (§82). The physical language, as opposed to purely mathematical languages, is characterized by two essentially distinct types of rules: *logical* rules represent the purely formal, nonempirical part of our scientific theory, whereas *physical* rules represent its material or empirical content. Moreover, this purely syntactical distinction between logical and physical rules is Carnap's precise explication for the traditional distinction between analytic and synthetic judgments (§§51–2). Finally, and what most concerns us here, it is also clear that these logical rules or analytic sentences of the language of science represent Carnap's precise explication for the constitutive – but nonsynthetic – a priori discussed above: these logical rules, in other words, syntactically represent Poincaré's (and Schlick's) conventions

18 This way of conceiving logical analysis is entirely impossible within the conception of logic of *Principia Mathematica*, where there is no distinction between object language and metalanguage and the language of logic is essentially interpreted. And it is for this reason that earlier writers such as Schlick and Wittgenstein characterized logical analysis as an activity rather than a doctrine – Carnap here explicitly rejects such "mysticism" (§73). Compare Goldfarb (1979), Ricketts (1985). (Parenthetical references in this paragraph are to section numbers of *Logical Syntax*.)

and Reichenbach's axioms of coordination.[19] Accordingly, although logical rules, just as much as physical rules, can indeed be revised in the progress of empirical inquiry, there is still a sharp and fundamental distinction, within the context of any *given* stage of inquiry, between the two types of rules.

Carnap's solutions to the two problems depicted above as lying at the basis of the logical positivists' radically new philosophical position are therefore as follows: the distinction between conventions or coordinative definitions and empirical laws properly so-called is just the distinction between logical and physical rules, analytic and synthetic sentences; the standpoint and method of philosophy – now conceived as logical analysis – is just the logical syntax of the language of science. Unfortunately, however, it proves to be impossible to implement both of these solutions simultaneously. More precisely, it proves impossible to implement both simultaneously together with what I take to be another linchpin of Carnap's distinctive philosophical stance: the claim to thoroughgoing philosophical neutrality. For it is a consequence of Gödel's incompleteness theorem that, for any language containing classical arithmetic among its logical rules or analytic sentences, the distinction between logical and physical rules can itself be drawn only within a meta-language essentially richer than classical arithmetic.[20] Implementing Carnap's analytic-synthetic distinction for such a classical language therefore results in a meta-language that, in particular, is in no way neutral between classical mathematics and intuitionism. It follows that there is no philosophically neutral metaperspective within which Carnap's distinctive version of conventionalism can be articulated coherently, and it is precisely here, it seems to me, that the ultimate failure of logical positivism is to be found.

Yet the implications of this failure for our contemporary, postpositivist philosophical situation have not, I think, been sufficiently appreciated. What I want to call attention to here are the very substantial parallels between central aspects of our postpositivist situation and basic elements of the positivists' own philosophical position. Thus, for example, it is now clear, I hope, that,

19 Thus, the first example of §50 parallels Reichenbach's account of the metric of physical space in *The Theory of Relativity and A Priori Knowledge*: it counts as "logical" in the flat space of classical physics but is "descriptive" (empirical) in the variably curved Riemannian space(-time) of general relativity. For further discussion of this parallel, see Chapters 3 and 4 (this volume).

20 For details, see Part Three, this volume. Carnap himself is perfectly clear about the technical situation here, but he fails to recognize its consequences for his claim to philosophical neutrality. Under Tarski's direct influence, he later abandoned the definition of analyticity – and indeed the program – of *Logical Syntax*.

far from being naive empiricists, the positivists in fact incorporated what we now call the theory ladenness of observation as central to their novel conception of science – a conception neither strictly empiricist nor strictly Kantian. Accordingly, they also explicitly recognized – and indeed emphasized – types of theoretical change having no straightforwardly rational or factual basis. In Carnap's hands, these conventionalist and pragmatic tendencies even gave rise to a very general version of philosophical "relativism" expressed in the *Principle of Tolerance.* If I am not mistaken, then, Cassirer's (1932/1951, p. 197) well-known characterization of the Romantic reaction against the Enlightenment – that the battle proceeded largely on the basis of weapons forged by the earlier movement itself – is perhaps even more true of the contemporary reaction against logical positivism. Since it proved ultimately impossible to combine all of the elements of positivist thought into a single coherent position, it would serve us very well indeed, I suggest, to become as clear as possible about the true character and origins of our own philosophical weapons.

PART ONE

GEOMETRY, RELATIVITY, AND CONVENTION

1

MORITZ SCHLICK'S *PHILOSOPHICAL PAPERS*

The appearance of these volumes[1] is an event to be welcomed by all students of twentieth-century philosophy of science and, indeed, by students of twentieth-century philosophy generally. Together they comprise the entire corpus of Schlick's published writings (and some previously unpublished writings) on epistemology, metaphysics, and philosophy of science with the exception of his magnum opus, *General Theory of Knowledge.*[2] Only Schlick's ethical writings fail to be represented exhaustively (and judging from some that are included, e.g., "On the Meaning of Life," vol. 2, pp. 112–29, this circumstance may not be cause for regret). Particularly noteworthy are translations of some of Schlick's best and most important, yet previously untranslated, papers (I have in mind especially "The Philosophical Significance of the Principle of Relativity" (1915), vol. 1, pp. 153–89, and "Experience, Cognition, Metaphysics" (1926), vol. 2, pp. 99–111) and the inclusion of Henry Brose's 1920 translation of *Space and Time in Contemporary Physics* (1917) (vol. 1, pp. 207–69), which has long been out of print. As a whole, we are presented with a fascinating, if somewhat chaotic picture of analytic philosophy in the making.

It must be admitted that Schlick's work has serious limitations, especially so for a twentieth-century thinker. His grasp of modern logic is imperfect at best (although he is acquainted with, and explicitly refers to, Russell's [1903] *Principles of Mathematics,* he devotes a considerable amount of energy in *General Theory of Knowledge* [§14] to arguing that all rigorous

1 Mulder and van de Velde-Schlick (1978–9); references are given parenthetically in the text by volume and page numbers.

2 Schlick (1918/1985); references are given parenthetically in the text by section numbers.

inferences in mathematics and science can be expressed as sequences of syllogisms in the mood Barbara). His understanding of the logicist tradition of Frege, Russell, and early Wittgenstein is even worse (he is hopelessly confused about the crucial question of mathematical induction [vol. 1, pp. 84–5] and propounds the disastrous conception, later popularized by Ayer, that all arithmetical propositions are in themselves tautological, whether or not arithmetic is somehow a part of logic [vol. 2, pp. 344–5]). Moreover, Schlick never developed the habit of formulating philosophical views and arguments with what we would call logical precision: he often reads more like a preanalytic philosopher than a contemporary of Russell and Carnap.[3]

Despite these limitations, however, Schlick's contribution is invaluable. For he has both a wide-ranging synthetic sense and a remarkable ability to get to the heart of a matter. He clearly perceives the broad outlines of the philosophical, physical, and mathematical currents whose convergence resulted in the development of logical positivism, and he struggles honestly, acutely, and courageously – if not always coherently – with the intellectual stresses and strains produced by this convergence. And, in this connection, the fact that he is to some extent a preanalytic philosopher is not a defect but a virtue. Reading Schlick yields an improved appreciation of the continuities between analytic philosophy and its immediate ancestors, which, under the spell of logical positivism's revolutionary rhetoric, we are all too liable to forget.

I

According to one popular picture, logical positivism began as an empiricist or verificationist movement in the tradition of Hume, Mach, and Russell's external world program. (See Schlick's own presentation of this picture in "The Vienna School and Traditional Philosophy" [1937], vol. 2, p. 495.) Add the "meaning-theoretic" orientation of Wittgenstein's *Tractatus*, and the verifiability theory of meaning is the result. However, if one reads the early (pre-1930) works of the positivists themselves, a very different and, I think, much more interesting picture emerges. The verificationism of

3 It is instructive to compare Carnap, Reichenbach, and Schlick in this regard. Reichenbach appears to have had a much better understanding of logic than did Schlick, and he does attempt to formulate his views and arguments with "logical precision." Nevertheless, Reichenbach does not really put modern logic to work in his philosophizing; even more so than Schlick, his primary technical orientation is toward physics. Only Carnap is able to grapple technically and philosophically with both modern logic and modern physics, and this is undoubtedly a central reason for his preeminence.

the positivists did not develop along a direct line from Hume and Mach via Russell and Wittgenstein. At least equally important is an evolution from German neo-Kantianism and neo-idealism via Hilbert and Einstein.[4] Schlick's writings on relativity theory provide a striking illustration of this evolution.

The essential background against which Schlick's work on relativity theory develops is the philosophical view expressed in *General Theory of Knowledge*.[5] That view diverges from stereotypical positivism or empiricism in two important respects. First, the conception of knowledge is explicitly "holistic." Knowledge is essentially a *system* of interconnected judgments whose concepts get their meaning from their mutual relationships within this system (§§3–9). Knowledge or cognition (*Erkennen*) is to be sharply distinguished from acquaintance (*Kennen*) or experience (*Erleben*) of the immediately given (§12). Direct confrontation with the given cannot yield knowledge; rather, knowledge always involves subsumption under concepts and, since concepts have meaning only in a system of judgments, always goes beyond the immediately given. Thus, for Schlick, unlike Russell, "*knowledge* by acquaintance" is an impossible contradiction. The paradigm of knowledge is not provided by Russellian sense-datum judgments but, for example, by Maxwell's equations – for these maximize the unified interconnectedness of our system of judgments (§11).

Second, Schlick militantly opposes the positivism of Mach and the phenomenalism of Russell's external world program (§26). Science deals with

4 The neo-Kantian and neo-idealist influence on the early positivists has been widely neglected – again, largely because of positivism's own anti-Kantian rhetoric. To get an initial appreciation of this influence, one has only to list some of the authors referred to by the two great works of the period: Schlick's *General Theory of Knowledge* and Carnap's (1928a) *Aufbau*. Names such as Cassirer, Driesch, B. Erdmann, Külpe, Husserl, Natorp, Vaihinger, and Wundt predominate. (As Alberto Coffa has emphasized to me, Schlick's own initiation into neo-Kantianism probably came from Helmholtz via Planck, who was the adviser for Schlick's doctoral dissertation.)

5 I do not want to mislead the reader by calling *General Theory of Knowledge* the "background" to Schlick's work on relativity. After all, the former was published in 1918, whereas the two major works on relativity appeared in 1915 and 1917, respectively. Nevertheless, *General Theory of Knowledge* reads like a prerelativistic piece of philosophy. It is striking that there are very few explicit references to Einstein; in addition, it seems quite clear that the paradigm of a physical theory is late nineteenth-century Maxwellian electrodynamics. Warren Goldfarb has suggested to me that Schlick had indeed incorporated relativity into his philosophy in *General Theory of Knowledge*, but he kept its influence veiled in order to "soften up" and avoid shocking his philosophical audience. I confess that this suggestion seems implausible to me in view of the militant anti-Machian stance of Schlick's book; I think it is more likely that he had simply done most of the thinking and writing for the book before he grappled philosophically with Einstein's new theories. But these are matters for further scholarship to decide. (See the Postscript to this chapter for further discussion.)

real unobservable entities – Schlick goes so far as to call them "transcendent" entities or "things-in-themselves" – which cannot be understood as mere logical constructions from sense data (Mach's "elements"). To think otherwise is to confuse knowledge and acquaintance, concepts (*Begriffe*) and images (*Vorstellungen*). While unobservable "transcendent" entities (atoms, electrons, the electromagnetic field) are not intuitable or even picturable, this does not prevent them from being conceptualizable and knowable. Knowledge and conceptualization do not require experience (*Erleben*) or intuitive representation (*Vorstellung*), but only a relation of *coordination* (*Zuordnung*) or *designation* (*Bezeichnung*) between concepts and objects (§10). Hence, although we cannot experience, intuit, or picture the entities of modern science, we can catch them in the net of our concepts – and this is all that knowledge or cognition requires. Of course, the net of our concepts must ultimately come in contact with experience or the given (in a way that Schlick never succeeds in making entirely clear), but this minimal empiricism would not be questioned by Kant, for example, or by contemporary scientific realists.

In other words, Schlick was not a positivist or strict empiricist in 1918, but a neo-Kantian or "critical" realist[6] – his viewpoint is perhaps best described as a form of "structural realism."[7] Why did Schlick (and therefore logical positivism generally) move away from this view? An answer emerges from Schlick's writings on relativity theory. Briefly, developments in modern mathematics and modern physics make a Kantian solution to the question of theory choice untenable. We are left with the problem of theoretical underdetermination in all its sharpness, and a "conventionalist" or verificationist solution becomes overwhelmingly tempting. Moreover, the development of Einstein's general theory of relativity appears to realize such a "conventionalist" solution completely: Machian empiricism appears victorious.

Recall the Kantian conception of scientific cognition. On the one hand, we have pure intuition, the a priori forms of space and time, in which all mathematical reasoning takes place via a process of "construction." On the other hand, we have the empirical or phenomenal world, which is nothing

6 I do not want to mislead the reader by calling Schlick a neo-Kantian. Schlick always rejected the synthetic a priori and Kant's theory of space and time. Yet Schlick's theory of *judgment* and *cognition* was Kantian (and antiempiricist) in its "holism" and "formalism." Schlick's view results from accepting a Kantian (or neo-Kantian) conception of judgment and cognition while rejecting Kant's doctrine of pure intuition.

7 As such, the view of *General Theory of Knowledge* has close affinities with Russell's (1927) *Analysis of Matter*, which also rejects phenomenalism in favor of "structural correspondence." One can get a partial sense of Schlick's view of this period from vol. 1, pp. 201–5.

but a distribution of matter (content) within the forms provided a priori by pure intuition. So we know a priori that the empirical world is subject to all the laws of pure mathematics and, because it also must conform to the unity of consciousness, to the continuity and conservation principles expressed in the Analogies of Experience as well. In particular, we know a priori: (i) space-time is Euclidean-Newtonian, (ii) Galilean kinematics, (iii) the law of inertia, (iv) $F = ma$, (v) any two pieces of matter (point-masses) are related by forces of attraction and repulsion (this last follows from the definition of matter developed in the *Metaphysical Foundations of Natural Science* [1786]). The only task left to a posteriori or empirical cognition, then, is a determination of the actual magnitudes of the forces specified in (v). For example, we apply (i)–(v) to the observed (Keplerian) orbits of the planets to "deduce from the phenomena" Newton's formula for gravitational force, which we then "make general by induction." The point is that no theoretical underdetermination infects this process. There is only the ordinary inductive uncertainty that always can be corrected by future observations. For Kant, scientific cognition has – in principle, anyway – a unique and determinate outcome.[8]

But developments in mathematics and physics upset this Kantian synthesis from all sides. Nineteenth-century foundational work, especially Weierstrass's "rigorization" of analysis and Hilbert's (1899) *Foundations of Geometry*, upsets Kant's conception of mathematical reasoning: "rigorous" inference has no need for intuition and proceeds purely conceptually.[9] So, pure mathematics has nothing in particular to do with space and time. By the same token, Einstein's work on special relativity upsets Kant's conception of physical reasoning. In particular, (i)–(v) are no longer "fixed points," but can themselves be revised in the course of scientific inquiry. And, if there are no such fixed points, cognition no longer has – even in principle – a determinate outcome: theoretical underdetermination is the result.

Schlick begins to wrestle with these problems in "The Philosophical Significance of the Principle of Relativity" (1915). In my opinion, this is one of his very finest papers: it is clear, penetrating, accurate, and balanced; even

8 For details of this reading of Kant's philosophy of science, see Friedman (1992a,b).

9 Although Schlick, for one, is far from clear about this, these nineteenth-century foundational developments require, for the first time, complicated patterns of polyadic reasoning (as is evident in the Cauchy–Bolzano–Weierstrass treatment of continuity, convergence, etc., and especially in Weierstrass's distinction between pointwise and uniform properties). It is no accident that Frege's work follows closely on the heels of that of Weierstrass. For Kant, on the other hand, logic is essentially monadic or syllogistic: Kantian "concepts" are monadic concepts. For details see Friedman (1992a, Chapters 1 and 2).

today, I know of no better philosophical introduction to special relativity. We are fortunate that it is now easily accessible to the English-speaking student. The aim of the paper is to examine the impact of Einstein's new theory on the two prevailing philosophical systems of the day: the neo-Kantianism of Cassirer, Natorp, and the Marburg school and the positivism of Mach and Petzoldt (vol. 1, pp. 153–5). As we shall see, neither system comes out unscathed; neither is able to assimilate fully the new physical discoveries.

After giving an extremely clear account of what he calls the principle of relativity – that all inertial frames are physically indistinguishable and no (uniform rectilinear) motion relative to the ether can be detected – Schlick goes on to point out that this *principle*, which is indeed a well-confirmed empirical law, does not yet amount to the *theory* of relativity (vol. 1, pp. 159–62). For there are two different ways of accommodating the empirical facts. We can, with Lorentz and Fitzgerald, maintain that (uniform rectilinear) motion relative to the ether exists, but remains undetectable because of compensatory contractions and retardations – thereby retaining Euclidean-Newtonian space-time and Galilean kinematics. Alternatively, we can, with Einstein, maintain that (uniform rectilinear) motion relative to the ether does not exist, and that indistinguishable reference frames are fully equivalent as well – thereby abandoning Euclidean-Newtonian space-time and Galilean kinematics. We have a choice here because the two theories are empirically equivalent: they are equally consistent with all experiential data (vol. 1, p. 164).

On a Kantian conception, there is, of course, no choice here. Our spatiotemporal intuition has a determinate and objective structure, and so, Euclidean-Newtonian space-time and Galilean kinematics are "fixed points" as above. Therefore, only the Lorentz-Fitzgerald theory is a live option. For Schlick, on the other hand, spatiotemporal intuition has a merely "subjective" and "psychological" character (vol. 1, pp. 162–3), and no such a priori considerations can rule out Einstein's theory. Moreover, Einstein's theory has important methodological advantages over the Lorentz-Fitzgerald theory. In particular, it possesses the kind of greater *simplicity* that has always moved scientists to prefer one theory over another – as in the choice of Copernican over Ptolemaic astronomy, for example (vol. 1, pp. 164, 170–1). Unfortunately, however, there is no way to justify our methodological preference for the simpler theory: we cannot argue that simpler theories come closer to the truth (vol. 1, pp. 169–70).

Schlick obviously has gotten himself into a quandary here. At times, he flirts with a "conventionalist" or verificationist solution. Thus, with a nod to Poincaré, we can maintain that two empirically equivalent theories are equally correct as well: there is no need for a choice between them (vol. 1, pp. 167–9). Yet, because Schlick is a convinced Einsteinian, he cannot remain satisfied with this way out. He proceeds to argue that perhaps simpler theories are closer to reality after all, since they contain fewer "*arbitrary* elements" (vol. 1, pp. 171–2). By the end of the paper, he is prepared to assert that the ether hypothesis is senseless and devoid of all "physical meaning" (vol. 1, p. 185). With respect to the real case of empirically equivalent theories that the development of relativity theory itself supplies, then, Schlick is not ultimately willing to opt for strict empiricism. He does not want to say that empirically equivalent theories are necessarily equally correct.

Moreover, as Schlick makes amply clear, relativity theory itself retains important elements of unobservable, theoretical structure. Although Einstein's theory does have fewer unobservable elements than the Lorentz-Fitzgerald ether theory, it does not eliminate such elements completely. In particular, while it is true that relativity theory eliminates absolute rest and velocity, it does not eliminate *absolute acceleration and rotation.* So, the kind of limitless relativity of motion maintained by Mach on empiricist grounds does not find expression in Einstein's theory (vol. 1, pp. 179–84). Furthermore, even if relativity theory were to realize Mach's ideas, this could be based only on actual empirical findings (like the empirical findings that in fact support the special or restricted principle of relativity). A purely philosophical commitment to strict empiricism is quite insufficient (vol. 1, p. 183).

So things stood in 1915. In 1916, however, Einstein brought years of work on a relativistic theory of gravitation to successful completion with the publication of "The Foundation of the General Theory of Relativity." Sections 1–3 of that paper make far-reaching philosophical claims on behalf of the new theory. In particular, Einstein claims finally to realize the thoroughgoing relativity of motion envisioned by Mach (hence the name of the new theory) and to remove from space and time "the last vestige of physical objectivity" (1916/1923, p. 117). The only spatiotemporal features left invariant under the *arbitrary* substitutions allowed by the principle of general covariance are space-time *coincidences*: meetings of material particles, matching of endpoints of rigid rods, coincidences between the hands of a clock and points on the dial, and so on. It follows that only such "observable" events are physically real. Abstract theoretical structures – assignments of "absolute" motion, attributions of a particular metrical geometry, and so on – can be arbitrarily transformed at will and are therefore only conventionally chosen aids for facilitating the description of the totality of space-time coincidences.

In *Space and Time in Contemporary Physics* (1917) (vol. 1, pp. 207–69), Schlick embraces these new ideas with easily understandable enthusiasm. Einstein's construction of an actual physical theory satisfying the principle of general covariance offers intoxicating relief from the skeptical problem of theoretical underdetermination. In particular, we are able to follow Poincaré in regarding empirically equivalent metrical geometries as equally correct (vol. 1, pp. 223–33) and to follow Leibniz and Mach in regarding all motion as essentially relative (vol. 1, pp. 233–43). We are left with a view that comes very close indeed to verificationism or strict empiricism:

> The adjustment and reading of all measuring instruments of whatsoever variety – whether they be provided with pointers or scales, angular-diversions, water-levels, mercury columns, or any other means – are always accomplished by observing space-time coincidences of two or more points. This is also true above all of apparatus used to measure time, familiarly termed clocks. Such coincidences are, therefore, strictly speaking, alone capable of being observed; and the whole of physics may be regarded as a quintessence of laws, according to which the occurrence of space-time coincidences takes place. Everything else in our world-picture which cannot be reduced to such coincidences is devoid of physical reality, and may just as well be replaced by something else. All world pictures which lead to the same laws for these point-coincidences are, from the point of view of physics, in every way equivalent. (vol. 1, p. 241)

Thus, the development of physical theory itself appears to make Machian positivism a much more viable position.[10]

Nevertheless, Schlick still draws back from a thoroughgoing positivism (after all, it is 1917, and *General Theory of Knowledge* is just being completed).

10 It is now more or less generally known that this positivistic interpretation of general relativity is completely untenable. General covariance implies neither a generalized relativity of motion nor a "conventionalist" conception of physical space (or space-time) as "metrically amorphous." As a matter of fact, any space-time – including the space-times of Newtonian physics and special relativity – can be given a generally covariant description. What distinguishes general relativity is the use of a nonflat (non-Euclidean) space-time of variable, mass-energy–dependent curvature. General relativistic space-time has a perfectly determinate and objective (although variable) metrical structure, and absolute acceleration and rotation have much the same status as they do in special relativity: Machian relativity is not realized. Furthermore, these (essentially mathematical) facts were established in Schlick's own time via the work of Kretschmann in 1917 and Weyl (1918). (The final clarification was provided by Cartan's great papers of 1923–4, which showed, in particular, how Newtonian gravitation theory also makes use of a nonflat space-time structure.) Yet the philosophical pressures leading toward a "conventionalist" misunderstanding of general relativity were quite irresistible (even to Einstein himself), and the notion of covariance is remarkably subtle and elusive. As a result, verificationist misinterpretations of general relativity have persisted almost to the present day. For details, and references, see Friedman (1983). For more on the incompatibility between general relativity and the "conventionalist" interpretation of metrical structure derived from Poincaré, see Chapter 4 (this volume).

Only space-time coincidences are real or objective, but not all space-time coincidences are literally observable. Real physical coincidences include such point-events as the collision of two elementary particles or the electromagnetic field strength taking on a particular value. Such point-events are not strictly observable, but they are "measurable," and this is all that is required for physical reality (vol. 1, pp. 264–6). The view we end up with appears to go something like this. Some pieces of theoretical structure – assignments of "absolute" motion, attributions of a particular metrical geometry – are given a positivist or "conventionalist" interpretation, but other pieces – atoms and electrons, the electromagnetic field – are given a "realist" interpretation. However, because Schlick provides no principled reasons for making this kind of distinction, his view is intrinsically unstable. Accordingly, his post-1917 writings take on an increasingly verificationist tone.

In his important critical study of Cassirer's work on relativity, "Critical or Empiricist Interpretation of Modern Physics?" (1921) (vol. 1, pp. 322–34), Schlick has nothing but praise for Mach (vol. 1, pp. 330–1). He even prefigures the principle of verifiability (vol. 1, p. 330): "*differences in reality may be assumed only where there are differences that can, in principle, be experienced*" and declares:

> if the principle is recognized and evaluated in its true significance, it can, I believe, be elevated to the supreme principle of all empirical philosophy, to the ultimate guideline which must govern our attitude to every question of detail, and whose ruthless application to all special problems is an exceedingly fruitful procedure. If this view is correct, the connection of relativity theory with empiricist theory of knowledge would then be seen as an intimate, strictly factual, and not merely external or contingent one. (vol. 1, p. 331)

(Of course Schlick's principle is not the principle of verifiability itself: it is a methodological principle, not a principle about meaning.) Finally, in a popular lecture from 1922, "The Theory of Relativity in Philosophy" (vol. 1, pp. 343–53), Schlick enunciates the principle, itself derived from Einstein's 1916 paper, that (vol. 1, p. 345) "*only something really observable should be introduced as a ground of explanation in science.*" (Note that this principle is much stronger than the above-mentioned principle from 1921.) He again sings Mach's praises (vol. 1, p. 347), and, when he criticizes "positivism," he now means only the "subjectivist" and "relativist" views of Petzoldt (vol. 1, pp. 347–8). We here stand on the very threshold of logical empiricism.[11]

11 It appears that Schlick's work on relativity was instrumental in "converting" the other early positivists to strict empiricism. Thus, Reichenbach's early work, *The Theory of Relativity and A Priori Knowledge* (1920), takes an explicitly antiempiricist and neo-Kantian line. In marked contrast to the well-known views of *The Philosophy of Space and Time* (1928), he argues that,

II

Schlick's lifelong struggle with the form/content distinction is of particular interest. In *General Theory of Knowledge,* his starting point is Hilbert's *Foundations of Geometry* and the notion of axiomatic or *implicit* definition (§7). According to the conception that Schlick derives from Hilbert, the primitive terms of geometry require no intuitive meaning or content. All we need to know about these primitives for the purposes of pure geometry are their mutual logical relationships set up explicitly in the axioms. Points, lines, and planes are any system of objects whatsoever that satisfy these axioms. Hence, pure geometry is intrinsically "uninterpreted" and has nothing at all to do with intuitive space.[12] Schlick generalizes this picture to all concepts and all of science. The meaning of all concepts ultimately is determined by their mutual logical relationships within a system of scientific judgments, not by the intuitive content we may happen to associate with them. To think otherwise is to confuse concepts (*Begriffe*) with images (*Vorstellungen*).[13]

In other words, concepts generally are individuated by their logical forms, by their logical "places" within a deductive system. Nevertheless, scientific as opposed to purely mathematical concepts have content as well: they designate (*bezeichnen*) real "qualities" in nature, which, if our system of judgments is true, have the same formal properties as the corresponding concepts – that is, the real "qualities" satisfy our (initially "uninterpreted") judgments.

although the choices of coordinate system and rest system are indeed arbitrary or conventional, the choice of a *metric* is not. Accordingly, he rejects Poincaré's "conventionalism" (see the first footnote to Reichenbach's book). Schlick criticizes Reichenbach in the abovementioned article on Cassirer (vol. 1, p. 333), and, by 1922, Reichenbach has capitulated: "conventionalism" is correct and Poincaré is vindicated (1922/1978, pp. 34–5, 38–9, 44). Similarly, although it would certainly be incorrect to describe Carnap's (1922) position in *Der Raum* as empiricist (his central conclusion is also neo-Kantian: *n*-dimensional *topology* is an a priori condition of the possibility of experience), he does present a very clear and rigorous defense of "conventionalism" with respect to *metrical* structure that is directly inspired by Schlick's account of the significance of general covariance in *Space and Time in Contemporary Physics* (Carnap, 1922, p. 83). For further details see Chapters 2, 3, and 4 (this volume).

12 Such a conception of pure geometry would make no sense at all to Kant, for whom even geometrical *reasoning* requires the a priori manifolds supplied by pure intuition. Geometrical thinking is necessarily spatial in the intuitive sense, and "uninterpreted *geometry*" is a contradiction in terms.

13 Similar views are very much in evidence throughout the period. I have in mind especially C. I. Lewis's *Mind and the World Order* (1929) and Russell's *Introduction to Mathematical Philosophy* (1919) and *Analysis of Matter* (1927). Carnap's *Aufbau* presents the logically most sophisticated version. For further discussion, see Chapter 5 (this volume).

But the point that Schlick is most concerned to stress remains this: Knowledge or cognition (*Erkennen*) relates only to such *formal* properties; content is never itself an object of knowledge but only (at most) of experience (*Erleben*) and acquaintance (*Kennen*). So, the fact that some particular real "quality" is unintuitable (e.g., the electromagnetic field intensity) presents no obstacle whatsoever to its knowability: all that is required is a grasp of its purely formal properties (§§9, 12, 26).

Yet this kind of view suffers from overwhelming difficulties and verges on incoherence, for the relationship between form and content, concepts and reality, is left hopelessly obscure. Schlick makes the point himself with characteristic honesty and acuteness:

> in implicit definition we have found a tool that makes possible completely determinate concepts and therefore rigorously exact thought. However, we require for this purpose a radical separation between concepts and intuition, thought and reality. To be sure, we place the two spheres one upon the other, but they appear to be absolutely unconnected, the bridges between them are demolished. (§7)

Schlick is perfectly correct. The Kantian bridge between thought and reality – namely, pure intuition – has indeed been demolished. Hence, if we persist in a "holistic" and "formalistic" account of knowledge and judgment, we are driven toward idealism and the coherence theory of truth. In particular, we will have a hard time distinguishing physical or empirical knowledge from pure mathematics, on the one hand, and from arbitrary coherent systems of metaphysics or myth, on the other. Either way it will be difficult to maintain the Kantian commitment to mathematical physics as a paradigm of knowledge.[14] To Schlick, of course, the coherence theory of truth is anathema. So, as we shall see, he is continually tempted to renounce his "holistic"

14 This is precisely the path taken by the "logical idealism" of Cassirer and the Marburg School. Cassirer is perfectly happy to throw away the autonomous role played by Kant's pure intuition, and equally happy to dethrone mathematical physics from its position of preeminence. All coherent "symbolic forms" – mathematics, physics, art, religion, myth, and so on – are equally respectable and, in particular, have their own characteristic standards of "objectivity." We can no longer equate the objective with the results of natural science. See, e.g., Cassirer (1921/1953, Chapter VII). (Note added in 1998: This now strikes me as misleading and exaggerated in two important respects. First, Cassirer's "philosophy of symbolic forms" is a later development not characteristic of the Marburg School in general. Second, although this later view of Cassirer's does indeed include the doctrine that the various symbolic forms – art, religion, myth, and so on – have their own characteristic standards of objectivity, Cassirer also maintains a self-consciously Hegelian "phenomenology of knowledge" according to which the abstract and mathematical symbolic forms of the exact sciences constitute the *highest* stage of knowledge: see especially Cassirer [1929/1957].)

conception of meaning in favor of an "atomistic" empiricist conception which views the intuitively given as the ultimate repository of meaning after all.

Schlick returns to the form/content distinction in his classical paper, "Experience, Cognition, Metaphysics" (1926) (vol. 2, pp. 99–111). In the meantime he has moved to Vienna (1922) and read the *Tractatus* and drafts of Carnap's *Aufbau.* He again articulates the distinction between experience (*Erleben*) and cognition (*Erkennen*). Cognition relates always to purely formal or "structural" features that are expressed in logical relationships or "implicit definitions." Experience, on the other hand, involves content: the actual "qualitative" features of things, such as the redness of a red surface. Moreover, he makes the rather startling claim that all traditional metaphysics is based on a confusion of form and content, knowledge and intuition: metaphysics is the self-contradictory search for "intuitive knowledge" (vol. 2, pp. 107–11). So far, the basic ideas are familiar from *General Theory of Knowledge,* but there are also several important novelties.

First, form is said to be not only the sole object of knowledge, it is also all that is *communicable.* Form is what can be expressed in language; content is essentially private and incommunicable. Not only does content elude our cognition, it forever eludes our expression as well. Hence, logical form, and logical form alone, is the basis for intersubjective communication (vol. 2, pp. 99–102). Of course, this view, which is a natural extension of the views in *General Theory of Knowledge,* immediately makes the form/content distinction itself unstatable. This is why there is no such distinction in the *Tractatus,* for example: the substance of the world (= the totality of objects) is both form *and* content (2.025).

Second, Schlick has abandoned the realism of *General Theory of Knowledge.* In fact, he articulates the characteristic claim of later logical positivism that the difference between realism and strict empiricism cannot itself be expressed or stated. It makes no difference at all whether the objects of physics are viewed as logical constructions from experience or as "independent realities." All that matters is that there are objects with such-and-such statable formal properties (vol. 2, pp. 103–4). Note that Schlick here simply assumes, without argument, that the objects of physics *can* be conceived as logical constructions from experience. Perhaps it is because, meanwhile, he has become convinced that they *must* be so conceived. In any case, the net effect of these two moves is clear. The form/content distinction now coincides with the intersubjective/private distinction: content is always internal and experienceable. In *General Theory of Knowledge,* by contrast, content includes both internal, intuitable "qualities" and external, unintuitable "qualities." So, in 1926, the gulf between form and content, thought and reality, has at least been diminished.

By 1929 the remaining gulf somehow has magically disappeared, and the transformation is complete. Schlick is now the militant prophet of logical positivism. He expounds the verifiability theory of meaning (apparently for the first time in 1930; see vol. 2, pp. 156–7) and begins to give the following, characteristically "atomistic" and empiricist, argument for it: all explanations of meaning typically consist in defining some words in terms of other words; but this process cannot continue indefinitely or move continually in a circle; therefore, we must eventually attach words directly to experience in acts of ostension, and all meaning ultimately resides in the given. Thus, in 1930 (vol. 2, pp. 157–9): "if, say, I state the meaning of my words by elucidatory propositions and definitions, and thus by means of new words, we have again to ask for the meanings of these other words, and so on. This process cannot continue indefinitely, and always terminates at last in mere factual indications, in demonstrations of what is meant." And in 1931 (vol. 2, p. 220): "All of our definitions must end by some demonstration, by some activity. There may be certain words at the meaning of which one may arrive by certain mental activities just as I can arrive at the signification of a word which denotes colour by showing the colour itself."

Schlick articulates his new position most clearly and explicitly in "Positivism and Realism" (1932a) (vol. 2, pp. 259–84). There he gives the following argument:

> But when do I understand a proposition? When I know the meaning of the words that occur in it? This can be explained by definitions. But in the definitions new words occur, whose meaning I also have to know in turn. The business of defining cannot go on indefinitely, so eventually we come to words whose meaning cannot again be described in a proposition; it has to be pointed out directly; the meaning of the word must ultimately be *shown*, it has to be given. This takes place through an act of pointing or showing, and what is shown must be given, since otherwise it cannot be pointed out by me .... The *meaning* of every proposition is ultimately determined by the given alone, and by absolutely nothing else. (vol. 2, p. 264)

Here, the "holistic" conception of meaning of *General Theory of Knowledge* and "Experience, Cognition, Metaphysics" has been turned completely on its head.

By now the contradiction has become too explicit to ignore. So Schlick takes up the problem again in "Form and Content" (vol. 2, pp. 284–369), written in the same year. He again claims (now obviously under the influence of the *Tractatus*) that logical form or structure is the basis for expression and communication. The content of the given is not itself expressible; rather, when we appear to communicate about the given, we are only expressing *its* structural properties (e.g., the logical structure of "color space"). In marked contrast to the above passage from "Positivism and Realism," he says (vol. 2, p. 298): "understanding and meaning are quite independent of Content and have nothing whatever to do with it." To be sure, he also defends the verifiability theory of meaning, and even gives the familiar "regress of definitions" argument for it (vol. 2, pp. 310–11). But he now realizes that this argument must be seriously qualified in the context of the "structuralism" of the logical form conception. The passage is worth quoting in full:

> The chief reason why it was so generally believed that all real knowledge must in some way culminate in immediate acquaintance or intuition lies in the fact that they seem to indicate the points where we must look for the ultimate meaning of all our words and symbols. A definition gives the meaning of a term by means of other words, these can again be defined by means of still other words, and so on, until we arrive at terms that no longer admit of verbal definition; the meaning of these must be given by direct acquaintance: one can learn the meaning of "joy" or "green" only by being joyful or seeing green. Thus the final understanding and interpretation of a proposition seems to be reached only in those acts of intuition – is it not through them, therefore, that the real knowledge which the proposition expresses is ultimately attained?
>
> The considerations in our first lecture have taught us already to what extent these remarks are true. We saw that our ordinary verbal language must be supplemented by pointing to objects and presenting them in order to make our words and sentences a useful means of communication, but we saw at the same time that in this way we were only explaining our language of words by a language of gestures, and that it would be a mistake to think that by this method our words were really linked to the content which intuition is supposed to provide for us. We showed that the meaning of our words was contained entirely in the *structure* of the intuitive content. So it is not true that the latter (the inexpressible greenness of green), which only intuition can furnish, actually enters into the understanding of knowledge. It cannot possibly do so. (vol. 2, pp. 321–2)

"Holism" has returned with a vengeance.

Yet Schlick still is not prepared to jettison "content" completely, although at times he comes very close to doing so (see, e.g., vol. 2, pp. 306–7). Why not? Unless our symbols have content as well as form, we are unable to distinguish applied sciences such as mathematical physics from "uninterpreted" deductive systems such as pure geometry:

> if we are to have a science of some domain of reality instead of a mere hypothetical-deductive system, then our symbols must stand for real content; for if they stood for mere structure, we should again in the end be left without meaning, for again there would be the possibility of many different interpretations. But actual science deals with reality, which is unique, and not with possibilities only, of which there are many.
>
> If this is the right answer it must appear difficult to reconcile with our former insight that content never enters into our propositions and that all expression is done solely by means of pure structure. (vol. 2, p. 331)

Such a reconciliation is difficult indeed! Here is Schlick's solution:

> the empty frame of a hypothetical-deductive system does have to be filled with content in order to become a science containing real knowledge, and this is done by observation (experience). But every observer fills in his own content. We cannot say that all the observers have the same content, and we cannot say that they have not – not because we are ignorant, but because there would be no sense in either assertion. (vol. 2, p. 334)

Plainly, Schlick has now reached the end of his rope.[15] How can subjective, private, and inexpressible content possibly help ground an intersubjective, public science?

III

After 1932, Schlick's intellectual creativity is attenuated considerably, and I will leave it to the reader to trace the twists and turns of thought in his final

15 Schlick is again ill-served by his poor understanding of the logicist tradition, especially as represented by Wittgenstein's (1922) *Tractatus.* From such a logicist point of view, the problem cannot even be set up. For, to formulate Schlick's problem, we must take a *meta-linguistic* stance toward the language of physics and regard the primitive terms of that language as uninterpreted *schematic letters.* But neither of these steps makes sense on a strict logicist conception. First, there is no distinction between object-language and meta-language. There is only one language: the single linguistic framework in which all our concepts and judgments are ultimately related. Second, there are no uninterpreted schematic letters. All symbols are either variables or constants (primitive or defined), and there is no room for a *choice* of interpretation. The interpretation of variables, for example, is fixed once and for all by their logical forms: first-level variables range over *all* individuals, second-level variables range over *all* functions of individuals, and so on. Similarly, there is no problem of distinguishing natural science (synthetic propositions) from pure mathematics (analytic propositions). On a logicist conception, pure mathematics is part of the logical framework of language itself, and so cannot possibly be confused with nonlogical propositions. Empirical science is just the totality of nonlogical (nontautologous) truths. (Of course, this conception of empirical science has nothing in particular to do with "experience" in the traditional sense, nor does it supply any kind of justification for mathematical physics as it now is or may become. Wittgenstein, for one, is quite sanguine about this: see

essays. (Essays 16, 18, and 19 of Volume 2 are especially interesting and revealing, and represent Schlick's side in the celebrated "protocol-sentence" debate with Carnap, Neurath, and Hempel – a debate about precisely the relation of "protocol-sentences," on the one side, to "experience," on the other.[16]) Instead, I would like to conclude with some general remarks about the significance of Schlick's work for our understanding of logical positivism and the broader philosophical context within which it develops. For it seems to me that careful attention to the actual history of logical positivism forces us drastically to revise our contemporary assessment of that movement, especially our contemporary picture of the relationship between "empiricism" and the new logic.

The standard picture of positivism goes something like this. Inspired by Hume, Mach, and Russell's external world program, the positivists adopted the concerns, problems, and ambitions of traditional empiricism more or less unchanged. Starting from a naive and "atomistic" conception of experience, observation, and the given, the problem was to show how the theoretical structures of physics could be grounded in or constructed from the given. Here modern logic supplied powerful new tools – the theory of relations and set theory – for carrying out this traditional project. These new tools were deployed with great ingenuity and resourcefulness – especially in Carnap's *Aufbau* – but, in the end, the dream of classical empiricism could not be realized: the objects of physics could not be defined or constructed from experience or the given. Hence, logical positivism is a failed philosophical movement, but its "logicization" of empiricism makes it a significant failure: it supplies a precise *proof* that strict empiricism cannot succeed.

Our reflections on Schlick and his philosophical context have shown, I hope, that this standard contemporary assessment is completely inadequate. The early positivists did not simply appropriate the new logic as a tool or technique for solving previously given philosophical problems. Rather, this logic

Tractatus 6.363–6.372.) Hence, if Schlick had really understood the *Tractatus*, he would never have become so hopelessly entangled with the form/content distinction. Nevertheless, we are fortunate that Schlick did misunderstand Wittgenstein, for *our* conception of logic is in many ways closer to Schlick's "Hilbertian" conception than to the logicist tradition (hence our contemporary problems about "inscrutability of reference" and "ontological relativity"). It would be wrong, however, simply to equate our conception of logic with Schlick's "Hilbertian" view. For our conception is the result of an evolution *within* the logicist tradition that took its present form only in the synthesis wrought by Gödel in 1930. See the excellent account in Goldfarb (1979). I am also indebted to Thomas Ricketts for very helpful conversations about these matters.

16 For details of this debate, see Uebel (1992) and Oberdan (1993). See also Chapter 6, §IV (this volume).

was itself the source of radically new problems and a radically transformed philosophical situation. In particular, the new logic made it possible to implement precisely an essentially Kantian "holistic" and "formalistic" theory of judgment and meaning, while at the same time dispensing with Kant's pure intuition and the synthetic a priori. In rejecting the synthetic a priori, the early positivists were indeed in agreement with traditional empiricism, but their "holistic" theory of judgment was a point of sharp disagreement. Our cognitive relation to experience or the given could not be understood on an "atomistic" paradigm, for all judgment – including perceptual judgment – makes sense only in the context of a total *system* of judgments. In other words, what we now call the "theory ladenness" of observation was actually a commonplace of early positivism (as well as of neo-Kantianism generally).

Yet the very factors that moved early positivism toward traditional empiricism and away from Kant – the rejection of pure intuition and the synthetic a priori – also made a genuine empiricist position problematic. Without pure intuition, the "formal" or "structural" basis for objective judgment – the infinitely rich set of logical forms of Frege's new logic – now has no particular connection with experience or the empirical world: objective judgment has no need for "content" in the Kantian sense. In this respect, it became much more difficult for the positivists to maintain a commitment to empiricism and empirical science than it was for Kant. It is clear, in any case, that empiricism cannot simply be combined with the new logic (as Russell attempted to do in his "logical atomism," for example); rather, the two stand in a kind of "dialectical opposition."

I believe that this dialectical opposition – exemplified so clearly and painfully in Schlick's struggles with the form/content distinction – animates both logical positivism and the analytic tradition as a whole. In Frege and the early Wittgenstein – thinkers whose primary concern was with mathematics and logic – the formal element predominates completely, and concern for empirical knowledge and epistemology in the traditional sense shrinks to the point of vanishing. In Schlick and Carnap – thinkers with an equal concern for physics and empirical science – the two elements are brought together with clearly evident resulting strains and tensions. (These tensions perhaps emerge most starkly and precisely in §§153–5 of the *Aufbau*, where Carnap is forced to argue that the notion of a "natural" or "experienceable" relation is itself a primitive concept of logic![17]) I believe that the "dialectical

17 For details, see Chapter 5, §III (this volume). See also the comments in the Postscript to this chapter (footnote 29).

opposition" between logic and experience, form and content, also informs Carnap's later work and the postpositivist philosophies of Quine and the later Wittgenstein. But these stories will have to wait for another day.

POSTSCRIPT: GENERAL RELATIVITY AND *GENERAL THEORY OF KNOWLEDGE*

Section I of this chapter presents an inadequate picture of the relationship between *General Theory of Knowledge* (1918), on the one hand, and Schlick's (1917) interpretation of Einstein's general theory of relativity, on the other. Indeed, footnote 5 explicitly suggests that general relativity had virtually no influence on *General Theory of Knowledge.* I now believe that this is quite wrong and that the influence of Einstein's theory on Schlick's epistemological treatise is in fact fundamental. In particular, a proper appreciation of this influence helps us to understand both the extent to which Schlick's "realism" of 1918 is actually supported by his 1917 interpretation of general relativity and the way in which Schlick thinks he can *restore* the bridge between thought and reality (§II, above) by making essential use of Einstein's theory. The key to both questions, it turns out, is an epistemological method, directly inspired by Einstein's theory, that Schlick calls "the method of coincidences."[18]

As explained in §II, above, Hilbert-style axiomatic systems, for Schlick, are paradigmatic of objective scientific conceptualization and objective scientific knowledge. And the sharp distinction, pertaining to such systems, between the formal-logical structure expressed in the axioms and their manifold possible interpretations (spatial, numerical, and so on) is mirrored in Schlick's central distinction between conceptual knowledge and intuitive acquaintance. Just as the Hilbertian focus on formal-logical structure is intended to purge geometrical deduction from possibly misleading reliance on spatial intuition, so as, in particular, to allow the logical relations of dependence between geometrical propositions to stand out more clearly, Schlick's theory of scientific conceptualization is intended to free it once

18 In what follows, I draw heavily on Friedman (1997). For the importance of the notion of "coincidence" in Einstein and Schlick (although from a somewhat different point of view), see Ryckman (1992). For the correspondence between Einstein and Schlick during this period, see Howard (1984). With respect to the issue of dating raised in footnote 5, this chapter, it is worth noting that the first (1918) edition of *General Theory of Knowledge* bears a dedication "to my dear father, on his seventieth birthday, 3 June 1916." This strongly suggests that *General Theory of Knowledge* was written more or less simultaneously with *Space and Time in Contemporary Physics.*

and for all from all vagaries of intuitive representation by allowing us to characterize scientific concepts in general solely in terms of their formal-logical relations to one another. In this way, the distinction between a formal axiom system for geometry (what we would now call an uninterpreted formal system), on the one side, and a possible interpretation for such a system via intuitive spatial forms, on the other, provides Schlick with the primary model for his own distinction between knowledge (*Erkennen*) and experience or acquaintance (*Erleben, Kennen*).

Moreover, the model of a Hilbert-style axiomatic system does indeed lead Schlick to the central problem of his early epistemology: elucidating the relation of such a system to the objects or realities that are now supposed to be known thereby. The problem is that a Hilbert-style axiomatic system, precisely in its purely formal-logical, essentially uninterpreted character, is deliberately and self-consciously divorced from all contact with reality:

> Implicit definition, by contrast [with concrete or ostensive definition], never stands in community or connection with reality, it denies this intentionally and in principle, it remains in the realm of concepts. A framework of truths constructed with the help of implicit definition never rests on the ground of reality, but, as it were, floats free, bearing, like the solar system, the guarantee of its stability within itself. None of the concepts appearing therein designate, in the theory, a real thing; rather, they mutually designate one another in such a way that the meaning of one concept consists in a determinate constellation of a number of the others. (1918/1985, §7, p. 37)

To now explain how knowledge of reality is possible – that is, in his own terms, how we can nevertheless set up a relation of designation or coordination between a Hilbert-style uninterpreted axiom system and some domain of real objects – is thus the *sine qua non* of Schlick's early scientific epistemology. His explanation, appropriately enough, is inspired by the second great advance in the foundations of geometry of his time, the application of non-Euclidean geometry to nature effected by Einstein's general theory of relativity.[19]

Reality, for Schlick, includes, paradigmatically, the domain of our private, immediately given data of consciousness, which constitute the totality of intuitive objects of acquaintance. These data are characterized by both

19 My neglect of this explanation in 1983 resulted from giving insufficient weight, in this connection, to Part Three of *General Theory of Knowledge*, which is entitled "Problems of Reality." The first section of Part Three (§22) begins: "Thus far we have left the realm of facts, of designated objects, wholly out of account, and we have occupied ourselves only with signs and the rules of their connection. . . . Now we step beyond this domain: we move from considering the form in which knowledge is presented to us to the content that is thereby presented; we turn away from the signs and towards the designated objects."

intuitive temporality and intuitive spatiality, in that there are immediately given temporal relations between them, and some of them (e.g., visual fields) exhibit intuitive spatial extendedness as well. Such intuitive spatiality and temporality are just as subjective as are the immediately given data of consciousness themselves, and, in this sense, it is perfectly correct to say that (intuitive) space and time are subjective (§§28–9). But this realm of intuitively given data is not the only reality, and the domain of intuitive spatiality and temporality, in particular, is not the only spatiotemporal reality. On the contrary, there is also an objective spatiotemporal reality, described, paradigmatically, by modern mathematical physics, which includes a great wealth of "transcendent" objects that are not intuitively given (electromagnetic fields and the like) and which extend far (in objective space and time) beyond the meager domain of realities immediately present to our consciousness. In this sense, the domain of immediate acquaintance is only a small fraction of existing reality.

Moreover, in this connection, especially, the failure sharply to distinguish between conceptual knowledge and intuitive acquaintance has produced serious philosophical confusion, and has led to the temptation, specifically, to restrict the domain of reality to the immediately given data of consciousness. Once we see, however, that knowledge means designation by concepts, and thus in no way requires intuitive acquaintance, we are in the best possible position definitively to resist this temptation:

> The intuition of things is not knowing and also not a precondition of knowing. The objects of knowledge must be *thinkable* without contradiction, that is, allow of a univocal designation via concepts, but they do not need to be intuitively representable. (§26, p. 232)

So the way is now open, in particular, to reject the subjective idealist "philosophy of immanence" on behalf of a fully robust scientific realism (§§25–6).

Indeed, Schlick's conception of knowledge as designation via concepts actually leads to an even stronger result; for it turns out that those objects which, in the first instance, are capable of such designation are not the intuitively spatial and temporal realities of immediate acquaintance, but rather precisely the objective "transcendent" realities described by modern mathematical physics as existing in objective, mathematical-physical space and time outside of our consciousness. It is precisely the latter realities, *rather than* the former, that constitute the proper objects of knowledge. This result already follows from Schlick's model of objective conceptual thought as given by mathematically precise axiomatic systems in which concepts are exactly specified, by means of implicit definitions, through their formal-logical relations to one another. (No such system is available, for example, for the domain of introspective psychology.) But it follows equally from Schlick's detailed explanation of how an abstractly specified system of implicit definitions acquires a relation of designation or coordination to the realities that are supposed to be known thereby, for this explanation yields a parallel and complementary emphasis on quantitative as opposed to qualitative knowledge.

Schlick explains how we set up the crucial relation of designation or coordination between our system of concepts and reality in §31, entitled "Quantitative and Qualitative Knowledge." We begin, to be sure, with the intuitive spatiotemporal ordering of the immediately given data of consciousness, for our construction of the objective or "transcendent" spatiotemporal ordering is based upon this subjective ordering:

> The ordering of our contents of consciousness in space and time is likewise the means by which we learn to determine the transcendent ordering of things outside our consciousness, and the latter ordering is the most important step towards their cognition.
>
> The problem now is to become clear how one proceeds from the intuitive spatiotemporal ordering to the construction of the transcendent ordering. This always occurs by the same method, which we can designate as the *method of coincidences*. It is epistemologically of the very highest importance. (§31, p. 272)

It turns out, however, that what is primarily knowable by this process is the quantitative structure of the "transcendent" ordering thereby effected. The qualitative structure of the immediately given data of consciousness with which we begin can itself only become known *after* we have fully articulated the objective ordering.

We construct the "transcendent" ordering, more specifically, on the basis of singularities or coincidences in our various intuitively given sensory fields. For example, I see the tip of my pencil touch my finger in my visual field and, at the same time, feel its touch on my finger in my tactile field. The intuitive spatiality of these two sensory fields is entirely different in the two cases, and they have, as such, no intuitive spatial relations to one another. I then bring them into relation by constructing a single, nonintuitive spatial ordering containing both the pencil and my finger, where a single point in objective space (the coincidence of my finger with the pencil tip) corresponds to both singular points in the two previously independent sensory fields. In this procedure, I abstract completely from the qualitative peculiarities of my sensory fields (color, tactile quality, and so on) and concentrate solely on their purely topological properties – the presence or absence of a singularity. And this focus on singularities or coincidences is also crucial from a scientific point of view, for it is precisely on the basis of such coincidences that the technique of numerical measurement now proceeds. We measure objective spatial intervals by observing the coincidences of the end-points of a measuring rod with points on a measured object; we measure objective temporal intervals by observing coincidences between events in a given natural process and pointer positions on a clock; and so on (§31, p. 275): "all measurement, from the most primitive to the most sophisticated, rests on the observation of spatio-temporal coincidences."

In the method of coincidences, then, I construct a numerical model, as it were, for an abstractly specified axiom system for mathematical physics[20] by carrying out measurements (of objective spatial and temporal intervals, but also of various objective physical magnitudes such as the electromagnetic field) based on my perceptions of measuring instruments and thus, in the end, on immediately given coincidences or singularities in my intuitive sensory fields. In this way, an abstractly specified axiom system acquires a relation of designation to quantitatively structured objective reality *by way of* the immediate data of consciousness, and the objective or "transcendent" spatiotemporal ordering of realities described by modern mathematical physics thereby becomes a genuine object of knowledge. It does not follow, however, that the purely qualitative data immediately present to consciousness themselves become objects of knowledge as well. On the contrary, precisely because they are not yet describable in truly quantitative fashion, such purely qualitative intuitive data are not yet objects of knowledge. They will only acquire this status, in fact, when they, too, are described in exact mathematical-physical fashion (§31, p. 288): "[t]he life of consciousness is thus only completely knowable in so far as we succeed in transforming introspective psychology into a physiological, natural-scientific psychology, ultimately into a physics of brain processes."[21]

We have not yet drawn a connection between the objective spatiotemporal ordering known via the method of coincidences and the new conceptions of space and time due to the general theory of relativity. Nor does Schlick himself make this connection explicit in the relevant parts of *General Theory of Knowledge.* Rather, it is in *Space and Time in Contemporary Physics,* written virtually simultaneously (footnote 18, above), that this crucial step in

20 Paradigmatic of such systems, for Schlick, are Maxwell's equations for the electromagnetic field and Einstein's field equation for gravitation (§27, pp. 242–3). In this way, such well-known slogans as "Maxwell's theory is Maxwell's equations," in the context of the Hilbertian tradition in geometry, constitute the immediate background to Schlick's epistemological conception.

21 This is the basis for Schlick's distinctive treatment of the mind-body problem in §§32–5.

Schlick's reasoning is explained. The most important chapter of the latter work, in this connection, is entitled "The General Postulate of Relativity and the Metrical Determination of the Space-Time Continuum," where Schlick draws a fundamental contrast between general relativity and both Newtonian physics and special relativity. In both of the latter two theories, he explains,

> [space] still preserved a certain objectivity, so long as it was still tacitly thought as equipped with completely determined metrical properties. In the older physics one based every measurement procedure, without hesitation, on the idea of a rigid rod, which possessed the same length at all times, no matter at which place and in which situation and environment it may be found, and, on the basis of this thought, all measurements were determined in accordance with the precepts of Euclidean geometry.... In this way, space was still left with a "Euclidean structure," as a separate and independent property, as it were, for the result of these metrical determinations was thought to be entirely independent of the physical conditions prevailing in space, e.g., of the distribution of bodies and their gravitational fields. (1917/1979, pp. 238–9)

But this is emphatically not the case in Einstein's new theory:

> If we want, therefore, to maintain the general postulate of relativity in physics, we must refrain from describing measurements and situational relations in the physical world with the help of Euclidean methods. However, it is not that, in place of Euclidean geometry, a determinate other geometry – e.g., Lobachevskian or Riemannian – would now have to be used for the whole of space, so that our space would be treated as pseudospherical or spherical, as mathematicians and philosophers are accustomed to imagine this. Rather, the most various kinds of metrical determinations are to be employed, in general, different ones at each position, and what they are now depends on the gravitational field at each place. (1917/1979, p. 240)

Space-time in general relativity now has no background geometry at all – neither Euclidean nor non-Euclidean – that would be determined independently of the distribution of matter therein; and, according to the general postulate of relativity (the principle of general covariance), the only background that remains is the topological or manifold structure of number quadruples, that is, the space-time coincidences, so that (p. 241): "the whole of physics can be conceived as a totality of laws in accordance with which the occurrence of these space-time coincidences takes place."[22]

22 In my original discussion in 1983, I was too quick to equate this passage with an endorsement of a version of verificationism, according to which only *observable* space-time coincidences are real; and a parallel error occurs in Friedman (1983, pp. 22–5).

In the final chapter, entitled "Relations to Philosophy," Schlick then explains the significance of Einstein's new view of space and time for epistemology. He points out that the objective spatial structure employed by physics is not intuitively given, but is rather a "*conceptual construction,*" that is, a "non-intuitive ordering, which we then call objective space and conceptually grasp through a manifold of numbers (coordinates)" (p. 260). Yet this objective conceptual construction proceeds, just as in *General Theory of Knowledge,* on the basis of the subjective spatiotemporal coincidences present in various sensory fields of various individuals:

> In order to fix a point in space, one must somehow, directly or indirectly, *point* to it..., that is, one establishes a spatio-temporal coincidence of two otherwise separate elements. And it now turns out that these coincidences always occur in agreement for all intuitive spaces of different senses and all individuals; precisely so is an objective "point," independent of individual experiences and valid for all, thereby defined. ... By closer consideration one easily finds that we attain to the construction of physical space and time exclusively through this method of coincidences and in no other way. The space-time manifold is nothing else than the totality of objective elements defined through this method. That it is precisely a four-dimensional manifold results from experience by the execution of this method itself.
>
> This is the result of the psychological-epistemological analysis of the concepts of space and time, and we see that we encounter precisely *that* meaning for space and time which Einstein has recognized as alone essential for physics, where he has shown it to best advantage. For he rejected the Newtonian concepts, which denied the origin we have described, and based physics instead on the concept of the coincidence of events. So here physical theory and epistemology extend their hands to one another in a beautiful alliance. (1917/1979, p. 262–3)[23]

The connection between Schlick's epistemological method of coincidences and the general theory of relativity could not be stated more clearly.

For Schlick, therefore, it is not only the case that Einstein's theory destroys the Kantian bridge between thought and reality – namely, pure intuition – but it also, via the method of coincidences, shows us how to restore such a bridge in a radically new form. General relativity is quite incompatible with pure intuition, of course, because space can no longer be conceived as "a separate and independent property ... entirely independent of the physical conditions prevailing in space." For precisely this reason, in fact,

23 The final sentence, which occurs in the original edition (1917, pp. 57–8), is omitted in the later edition, which instead adds some comments on the "paradoxes" of relativity.

the measurement procedures of "the older physics," in which space-time coordinates have a direct metrical significance, are now entirely inapplicable; and Einstein himself reacts to this situation by introducing a new conception of space-time coordinates having merely topological significance.[24] Schlick then generalizes Einstein's procedure in a most insightful way in his method of coincidences. A merely topological assignment of space-time coordinates can be taken as our model for how thought is related to reality in general: we project topological singularities in our subjective sensory fields onto a numerical assignment of both space-time coordinates and physical magnitudes constituting an objective interpretation for our "Hilbertian," initially uninterpreted, geometrical-physical theory.

Schlick's understanding of general relativity is thus intimately intertwined with a version – a sophisticated and abstract version, to be sure – of a classical "causal realist" theory of perception and knowledge. On one side are the intuitive realities of acquaintance directly given to our consciousness; on the other side are the "transcendent" realities in objective space and time. The latter are not given to our consciousness; the former are not (at least not yet) objects of conceptualization and knowledge. We can, nevertheless, set up a relation of coordination or correspondence between the two sides via the method of coincidences, in virtue of which, at the same time, our abstract conceptual systems obtain a relation of designation or coordination to the "transcendent" realities in objective space and time outside of our consciousness. It is thereby possible to have knowledge of these realities in spite of their "transcendence." Indeed, it is these realities, and *not* the "immanent" realities of consciousness, that are the proper objects of (conceptual) knowledge.

What then becomes of the argument of §I, above, according to which Schlick's assimilation of relativity theory was instrumental in converting him from "realism" to verificationism? We now see, in the first place, that this argument must be seriously qualified. For Schlick in fact combines "realism" and general relativity in a most ingenious way in the period of *General Theory of Knowledge.* Indeed, in view of the fact that Schlick still maintains a strong commitment to scientific realism – and a rejection, in particular, of the Machian "philosophy of immanence" – in the second (1925) edition of *General Theory of Knowledge,* we must admit that the more standard story,

24 This train of thought leading to general covariance is explained in detail by Einstein (1917/1920, §§23–8). It is closely related, moreover, to the opening pages of Einstein (1921/1923), which expound a sharp distinction between "pure axiomatic geometry" and "practical geometry" explicitly indebted to Schlick's *General Theory of Knowledge.*

according to which Schlick only adopted verificationism in the period of the Vienna Circle, under the direct influence of Wittgenstein's *Tractatus* and Carnap's *Aufbau*, is basically correct.[25] Nevertheless, and in the second place, we must also admit that there were indeed significant pressures pushing in a strongly verificationist direction arising from the development of relativity theory and its philosophical assimilation – pressures clearly evident in Einstein's own rhetoric.[26] Schlick resonates to this rhetoric in the passage quoted earlier (1917/1979, p. 241), and the remainder of §I does in fact give a correct representation of the increasingly Machian and verificationist tone of Schlick's writings on relativity theory. In this sense, Schlick's philosophical interpretation of relativity had a variety of levels and, in particular, contained the seeds of his later verificationism in a latent form. Although the influence of Wittgenstein and Carnap within the Vienna Circle was needed fully to complete the transformation, Schlick's assimilation of relativity involved quite independent motivations pushing in the same direction.

We should also note, finally, that Schlick's "causal realist" theory of perception and knowledge still harbors significant strains and tensions – precisely the tensions, in fact, that lead to his struggles with the form/content distinction discussed in §II, above. For the heart of this theory is the idea that private and subjective, intuitively given spatiotemporal relations constitute the basis for our construction of the objective and "transcendent" realm of nonintuitive spatiotemporal realities, which is thereby conceptually known. Yet the intuitive data of consciousness themselves are not (or at least not yet) conceptually known; we have (so far) no exactly specifiable system of judgments by which *they* can be designated. Even if we limit ourselves, as Schlick does, to the purely topological features of such intuitive data (the presence or absence of a singularity in some or another sensory field), Schlick's starkly dualistic opposition between intuitive acquaintance and conceptual knowledge presents us with a fundamental obscurity, if not an outright incoherence. How can initial data that are not (yet) conceptualized possibly serve as the basis for an epistemological construction of objective conceptual knowledge?[27]

25 For this kind of reading, see, e.g., Kraft (1950/1953). For further discussion, see Chapter 6, §IV (this volume).

26 See, e.g., Einstein (1916/1923, §§1–3); (1921/1923).

27 In his comments on Helmholtz's theory of spatiality and geometry in Hertz and Schlick (1921/1977, p. 172, n. 33), Schlick characterizes intuitive or psychological space as a purely qualitative "extendedness" provided by "that indescribable psychological moment of spatiality adhering to sense perceptions." In this passage (which is admittedly later than *General Theory of Knowledge*), Schlick explicitly suggests that intuitive space is wholly "indescribable" or ineffable. For further discussion of Schlick's appropriation of Helmholtz, see Friedman (1997).

Carnap (1928a) presents an epistemological construction of the objective physical world from the sensory fields (especially the visual fields) of an initial subject that is closely analogous to Schlick's method of coincidences. The heart of Carnap's procedure, in particular, is the projection of points of the visual field onto purely formal, purely numerical space-time coordinates in *n*-dimensional real-number space. But Carnap, unlike Schlick, does not take the sensory fields constituting the basis of this procedure as simply given. On the contrary, the domain of subjective experience itself – the domain Carnap calls the "autopsychological" – is subject to an elaborate conceptual construction in which a single initial nonlogical primitive (the relation of recollection of similarity *Rs*) is successively elaborated in the type hierarchy of *Principia Mathematica* so as to provide explicit logical definitions of a variety of notions pertaining to introspective psychology, including, in the end, the notion of the visual field itself. Here, in particular, Carnap ingeniously deploys the topological definition of dimension number recently formulated by Karl Menger to show that purely topological features of the "autopsychological" realm are just as formally – and therefore just as "objectively" – characterizable as are the metrical structures of physical space and time.[28] For Carnap, then, there is no dualistic opposition between intuitive experience and conceptual knowledge, and there is thus no problem, in particular, of somehow attaching conceptual knowledge, at its base, to nonconceptual intuitive content. All stages in the epistemological construction of knowledge take place *within* a single "constitutional system," which is equally formal-logical (and therefore equally "objective") at every level.[29]

28 For details of Carnap's epistemological construction, and the significance of the resulting conception of "objectivity," see Part Two, this volume.

29 As suggested above (see footnote 17, this chapter), Carnap does encounter a difficulty closely analogous to Schlick's problematic of "attaching" conceptual knowledge to nonconceptual intuitive content in §§153–5 of the *Aufbau*, a matter that is discussed in detail in Chapter 5, §III (this volume). Nevertheless, in light of the fundamental differences, just noted, between Carnap's procedure and Schlick's, I no longer think that this difficulty is such a serious one.

2

CARNAP AND WEYL ON THE FOUNDATIONS OF GEOMETRY AND RELATIVITY THEORY

At the end of the nineteenth century and, even more, in the early years of the twentieth century, the philosophy of geometry experienced unprecedented pressures and tensions. For the revolutionary new developments in the mathematical foundations of geometry and, even more, the application of many of these new mathematical ideas to nature in Einstein's theory of relativity seemed to suggest irresistibly that all earlier attempts to comprehend philosophically the relationship between geometry on the one hand and our experience of nature on the other were radically mistaken. In particular, the Kantian understanding of this relationship – according to which, geometry functions as an a priori "transcendental condition" of the possibility of our scientific experience of nature, and space is viewed correspondingly as a "pure form of our sensible intuition" – seemed to be wholly undermined by the new mathematical-physical developments. The question then – for philosophers, mathematicians, and physicists alike – was what new understanding of the relationship between geometry on the one hand and our experience of nature on the other was to be put in its place.

The variety of mutually incompatible answers that were given to this question is remarkable, in that precisely this variety reflects the true complexity – the manifold pressures and tensions – engendered by the radically new philosophical, mathematical, and physical situation. It is especially remarkable, in particular, how seldom a straightforwardly empiricist understanding of the relationship between geometry and experience – according to which geometry is an empirical theory like any other whose validity is straightforwardly verified or falsified by experience – was represented. Pure mathematician and logicians, for example, who were most concerned to reject the Kantian conception that geometrical reasoning requires "construction in pure intuition" (so that pure geometry is a synthetic science), argued that geometry is first and foremost a purely formal or analytic science having no intrinsic relation to experience whatsoever. Neo-Kantian philosophers, on the other hand, argued that, whereas the new mathematical-physical developments indeed undermined a strictly Kantian conception of the intuitive certainty and experience-constituting character of Euclidean geometry, in particular, Kant's most basic and fundamental insight into the "transcendental" function of space and geometry within physics necessarily remained valid. Indeed, even for more physically and empirically oriented thinkers such as Helmholtz, it seemed that there must be *some* important respect in which Kant's conception of space and/or geometry as presupposed, rather than supported or refuted, by empirical physics remains true. It still seemed that there must be something to the idea that we must first ascribe or contribute geometrical structure to nature before we can extract empirical mathematical laws from our experience of nature. Poincaré's conception of geometry as a free stipulation or convention was simply one of the forms in which this fundamental Kantian idea was preserved.

I

Rudolf Carnap's (1922) doctoral dissertation is a particularly interesting attempt to inject order into this rather chaotic situation. The dissertation was written at Jena, where Carnap studied the new mathematical logic with Frege, philosophy with the neo-Kantian Bruno Bauch, and also experimental and theoretical physics. After first attempting to write a dissertation in the physics department on the axiomatic basis of relativity theory (which the physics department found too philosophical), Carnap ended up receiving his doctorate under Bauch in the philosophy department. In the resulting dissertation, Carnap attempts to resolve the contemporary conflicts and tensions in the foundations of geometry – conflicts involving mathematicians, (neo-Kantian) philosophers, and physicists alike – by carefully distinguishing among three distinct types of space: *formal, intuitive,* and *physical.* Carnap (1922, p. 64) argues that the different parties involved in the various mathematical, philosophical, and physical disputes are in fact referring to different types of space, and, in this way, there is really no contradiction after all: "[all] parties were correct and could have easily been reconciled if clarity had prevailed concerning the three different meanings of space."

Carnap's conclusion is that his second type of space – intuitive space – is synthetic a priori and experience-constituting in precisely Kant's original sense. It is just that we need a more general structure than Kant's original three dimensional Euclidean space:

> It has already been explained more than once, from both mathematical and philosophical points of view, that Kant's contention concerning the significance of space for experience is not shaken by the theory of non-Euclidean spaces, but must be transferred from the three dimensional Euclidean structure, which was alone known to him, to a more general structure. However, to the question of what this latter is now to be, the answers are partly indeterminate, in that only isolated characteristics of the three dimensional Euclidean structure are proposed as requiring generalization, and partly contradictory, chiefly because of a failure to distinguish the different meanings of space and insufficient clarity about the conceptual relationship of the space-types themselves – especially the relation of the metrical to the superordinate topological ones. According to the foregoing reflections, the Kantian conception must be accepted. And, indeed, the spatial structure possessing experience-constituting significance (in place of that supposed by Kant) can be precisely specified as topological intuitive space with indefinitely many dimensions. We thereby declare, not only the determinations of this structuure, but at the same time those of its form of order [*n*-dimensional topological *formal* space] to be conditions of the possibility of any object of experience whatsoever. (1922, p. 67)

But how are we to understand this remarkable conclusion?

Formal space, for Carnap, is an abstract relational structure whose initially uninterpreted primitive terms receive a specifically spatial interpretation in intuitive space. And it is here, in fact, that the properly spatial character of space – that is, its intuitive *spatiality* – is distinguished from all other relational systems having the same formal structure. This intuitive interpretation of formal or purely mathematical space is not yet a physical or empirical interpretation, however. We obtain the latter only by taking the further step, in Carnap's language, of "subordinating [*Unterordnung, Subsumtion*]" actually experienced physical phenomena to the synthetic a priori form of intuitive space. (By contrast, Carnap calls the more familiar relation between formal and intuitive space "specification [*Einsetzung, Substitution*].") Carnap sums up the distinction between these two different "application" or "interpretation" relations thusly:

> The relation of [formal space] to [intuitive space] is that of the species of structures with determinate order-properties but undetermined objects to a structure with these same order-properties but determinate objects – viz., intuitively spatial forms. The relation of [intuitive space] to [physical space] is that of a form of intuition to a structure with this form made up of real objects of experience. (1922, p. 61)

There is no doubt, then, that Carnap intends his notion of intuitive space to be a generalized interpretation – appropriate to the new mathematical, philosophical, and physical context – of Kant's conception of space as an a priori form of intuition.

Carnap explicitly models his notion of spatial intuition on Edmund Husserl's concept of "essential insight [*Wesenserschauung*]," as developed especially by Husserl (1913/1931). As explained by Husserl, *Wesenserschauung* functions as follows. Just as in sense perception we are immediately presented with or immediately given a sensible particular or individual (a particular color spot, a particular tone, an individual spatial figure), so in *Wesenserschauung* we can immediately grasp the universal features that such given sensible particulars exemplify (the general color, the general tone, the general spatial figure). Because *Wesenserschauung* is thus directed at universal features rather than particular individuals, it is entirely independent of the particular individuals that actually exist in the real world. It can, for example, function just as well with an imagined individual or object of fantasy as with a real or actual individual. Indeed, for certain purposes, as in pure geometry, for example, *Wesenserschauung* functions even better with purely imagined individuals. And the key conclusion is now this: Because *Wesenserschauung* is independent of the particular individuals that actually exist in the real empirical world, it is a source of a priori rather than empirical knowledge. Thus, for example, through *Wesenserschauung* we obtain a priori knowledge of the structure of color space, of tone space, or (in pure geometry) of pure geometrical space. This knowledge is also synthetic a priori rather than analytic a priori, for it does not follows from the most general truths of pure formal logic alone. Pure formal logic, for Husserl, expresses the "essence [*Wesen, Eidos*]" common to all objects of thought whatsoever. The a priori "eidetic sciences" of color, of tone, or of geometrical space, however, hold only for particular subspecies of such objects.

Carnap himself explains the idea as follows:

> [H]ere, as Husserl has shown, we are certainly not dealing with facts in the sense of experiential reality, but rather with the essence ("Eidos") of certain data which can already be grasped in its particular nature by being given in a single instance. Thus, just as I can establish in only a single perception – or even mere imagination – of three particular colors, dark green, blue, and red, that the first is by its nature more akin to the second than to the third, so I find by imagining spatial forms that several curves pass through two points, that on each such curve still more points lie, that a simple line-segment, but not a surface-element, is divided into two pieces by any point lying on it, and so on. Because we are not focussing here on the individual fact – shade of color seen

> here-now – but on its atemporal nature, its "essence," it is important to distinguish this mode of apprehension from intuition in the narrower sense, which is focussed on the fact itself, by calling it "essential insight [*Wesenserschauung*]" (Husserl). In general, however, the term "intuition" may also include essential insight, since it is already used in this wider sense since Kant. (1922, pp. 22–3)

Carnap's debt to Husserl is therefore clear.

But which exactly are the truths of geometry revealed by *Wesenserschauung*? Husserl himself seems to have in mind the totality of truths of Euclidean geometry – as presented in Hilbert's well-known axiomatization, for example. For Carnap, by contrast, the whole problem, as it were, is to adapt Kant's notion of a form of intuition to the general theory of relativity. And here Carnap introduces a most ingenious idea. Euclidean geometry is indeed uniquely presented to us in intuition – more precisely, through Husserlian *Wesenserschauung*. But this very Euclidean intuition is valid only in small or limited spatial regions, so that we are in no way intuitively given, for example, the validity of the parallel postulate in global space. Rather, our intuition tells us only that the Euclidean axioms – which, for Carnap, are definitively given by Hilbert – are satisfied "in the smallest parts" of space in Riemann's sense. Intuition tells us, that is, that space is *infinitesimally* Euclidean; and it is this proposition that expresses the synthetic a priori knowledge characterizing intuitive space.

The point of physical space, for Carnap, is then to order and arrange the objects of our actual experience of nature in the intuitive space we have already constructed completely a priori. This makes sense, for Carnap, because the Husserlian *Wesenserschauung* yielding the basis of our a priori knowledge of intuitive space arises, in the first instance, from our sensible experience of actual spatial natural objects. By considering this experience in its general, essential, or "eidetic" aspects we then arrive at a priori laws governing the structure of intuitive space. Physical space – the space of physical theory – is much more than a mere aggregate of particular intuitively spatial experiences, however; it is rather a precise and consistent ordering of spatial objects in a single mathematical structure. And the point of such a structure is to assign mathematically precise spatial relationships to natural objects – determinate mutual angles and distances, for example – so that precise mathematical laws of nature then can be formulated.

Here, however, a problem arises for Carnap. Intuitive space, as he understands it, has only what he calls topological structure. More precisely, it has only infinitesimally Euclidean structure, so that we are given a priori, as it were, only the entire class of all possible Riemannian manifolds. But the kind of mathematical structure required by physical space is a full (local and global) metrical structure. More precisely, one particular Riemannian manifold must somehow be singled out. How is this to be done? For Carnap, such metrical structure is introduced by a freely stipulated convention. We can, for example, stipulate the Riemannian structure in question directly (as Euclidean, say, in the context of prerelativistic physics or via the Schwarzschild solution, say, in the context of the general theory of relativity). Alternatively, we can, if we prefer, begin from what Carnap calls a "measure-stipulation [*Maßsetzung*]" – roughly, the stipulation that a particular physical body is rigid – and then indirectly determine the metrical structure through measurement. The important point, for Carnap, is that, on either alternative, the particular metrical structure one arrives at is in no way inherent in the actual empirical facts of nature; it expresses instead merely the outcome of our own free choice – a choice, to be sure, that is nonetheless absolutely essential, for otherwise we simply could not make precise mathematical determinations at all.

But why exactly is metrical structure thus independent of the actual empirical facts of nature? Carnap explains this idea through the concept of a "matter of fact [*Tatbestand*]" of experience. We know, from our consideration of the *Wesenserschauung* underlying intuitive space, that the natural objects actually given to us in experience (or even in mere imagination) have certain necessary or a priori spatial features – just those necessary features expressed in the structure of intuitive space. These – and only these – are the spatial features that empirically given natural objects have according to their very nature, as it were. These formal features – and only these – belong to what Carnap calls the "necessary form" of spatial objects of experience. But, as we have seen, the formal features in question are exhausted by what Carnap calls topological structure; no metrical structure (beyond the *infinitesimal* metrical structure) is in fact to be found here. Such further (metrical) spatial structure is thus not inherent in the actual given facts – the "matters of fact" – of experience according to their (spatial) nature, and must instead be conventionally imposed on these facts by us in the form of a freely chosen stipulation:

> Now, we have called experience, in so far as it is presented only in the uniquely necessary form that contains no freely chosen stipulations whatsoever, "matter of fact [*Tatbestand*]." Therefore, only the spatial determinations contained in matters of fact can be conditions of the possibility of experience. And these, as we have seen, are only the topological, but not the projective and above all not the metrical relations. (1922, p. 65)

Carnap's metrical conventionalism here is therefore entirely unique and should not be assimilated to any of the other then-current forms of conventionalism: that of Dingler, Poincaré, Schlick, or Reichenbach. For Carnap's version rests in the end on his own peculiarly hybrid conception of space as a form of intuition – a form of intuition necessarily lacking a full (local and global) metrical structure.

II

We can gain a deeper understanding of Carnap's position by juxtaposing it with the, in important respects, quite similar position developed at roughly the same time by Hermann Weyl. Weyl was, of course, one of the most penetrating and remarkable contributors to the debate focused on the new situation in the foundations of geometry and the theory of relativity – especially remarkable because he contributed to the mathematical, physical, and philosophical aspects of this debate. It appears, moreover, that Weyl's work was centrally important to Carnap's thinking in his dissertation. Weyl (1918/1922) is recommended "in the first place" as a reference for the general theory of relativity, and, for the idea that Euclidean geometry is valid in the small (particularly in the context of general relativity), the reader is referred to Weyl's *Erläuterungen* to Riemann (1919): Carnap (1922, pp. 81, 84). It is also worth noting, in addition, that Carnap (1922) is one of the very few philosophical works in the literature on general relativity cited in the fifth (1922) edition of Weyl's book (note 1 to chapter IV) – along with Schlick (1917/1920) and Cassirer (1921/1923).

The similarities between Carnap's conception and Weyl's are striking indeed. For Weyl also articulates a position according to which space is "essentially" or by its very "nature" *infinitesimally* Euclidean, but not, of course, either *locally* or *globally* Euclidean. Thus, according to Weyl, what he calls the *nature* of space is given by the circumstance that at each point of the manifold the tangent space bears the same Euclidean metrical structure. What Weyl calls the mutual *orientation* of the metrical structure as we move from point to point, from tangent space to tangent space, is, by contrast, entirely accidental and contingent:

> The *nature* of the metric characterizes the a priori essence [*Wesen*] of space in regard to the metrical; it is *one*, and it is therefore also absolutely determined and does not participate in the unavoidable vagueness of that which occupies a variable position in a continuous scale. What is not determined through the essence of space but rather a posteriori – i.e., accidental and capable in itself of free and arbitrary virtual variations – is the mutual *orientation* of the metrics at different points; in reality it stands in causal dependence with matter and can – participating in the vagueness of continuously variable magnitudes – never

> be fixed exactly in a rational way, but always only via approximation and also never without the help of immediately intuitive references to realtity. One sees that the Riemannian conception does not deny the existence of an a priori element in the structure of space; it is only that the boundary between the a priori and the a posteriori has been shifted. (1918/1922 [5th ed.], §13)

And it is clear, moreover, that Weyl, like Carnap, conceives this new Riemannian conception of the spatial a priori as a generalization of the Kantian conception – a generalization adapted to the new scientific situation created by relativity theory.[1]

The kinship between the two thinkers becomes especially intriguing when we note that Weyl, like Carnap, bears a substantial philosophical debt to Edmund Husserl. Thus, in the Introduction to Weyl (1918/1922), although Weyl explicitly states that he will refrain from entering more deeply into the purely philosophical aspects of the problem in the present book, he nonetheless adds several paragraphs of a self-consciously philosophical nature in which he provides a general outline of the relationship between mathematical-physical theorizing and the realm of immediate subjective experience – the realm of phenomenological "pure consciousness [*reine Bewußtsein*]." An endnote to these philosophical paragraphs then explains that "the precise version of these thoughts depends most closely on Husserl's *Ideen zu einer reinen Phänomenologie*" (1918/1922, note 1 to the Introduction). Moreover, at the end of Chapter II of the fourth edition, at the conclusion of his own presentation of purely infinitesimal Euclidean geometry as constituting the essence or nature of space, Weyl explicitly connects his analysis with Husserlian phenomenology in an even more striking fashion (1918/1922, §18): "The investigations undertaken in Chapter II concerning space seem to me to be a good example of the essential analysis [*Wesensanalyse*] striven after by the phenomenological philosophy (Husserl)." Could it be, then, that Weyl's underlying (if not explicitly expressed) philosophical conception is also close to Carnap's?

That this is not the case emerges when we reflect more closely on how Weyl arrives at the idea that infinitesimally Euclidean geometry expresses the a priori essence or nature of space. Does Weyl, like Carnap, arrive at this idea by considering our immediate, quasi-perceptual intuitive consciousness of very small spatial regions? Not at all; he instead presents a subtle and

1 Although Weyl does not explicitly mention Kant in the above-cited passage, closely parallel passages in other works make the Kantian context perfectly clear. See especially Weyl (1922, p. 116), where the infinitesimally Euclidean nature of space is said to be "*characteristic of space as form of appearance*," and Weyl (1927/1949, §18), where Kant is explicitly discussed.

extremely complex mathematical analysis that is intended to generalize the group-theoretic solution to the "space problem" of Helmholtz and Lie to manifolds of variable curvature. Just as Helmholtz and Lie derived the Pythagorean (that is, infinitesimally Euclidean) character of the metric from a postulate of free mobility – which, however, thereby limited their solution to Riemannian manifolds of constant curvature – Weyl derives the Pythagorean character of the metric from a more general group-theoretic argument of a strictly infinitesimal character. Weyl's more general group-theoretic conditions thus yield the Pythagorean (i.e., infinitesimally Euclidean) nature of the metric without restricting us to homogeneous spaces.[2] The important point for our purposes, however, is that Weyl's deep mathematical analysis of the space problem – an analysis he was only able to bring to successful fruition by intensive efforts over several years – has almost nothing in common with Carnap's conception of our a priori knowledge of space.

This stands out particularly clearly, in fact, in the continuation of the passage, cited above, in which Weyl compares his analysis of space to Husserlian *Wesensanalyse*:

> The investigations undertaken in Chapter II concerning space seem to me to be a good example of the essential analysis [*Wesensanalyse*] striven after by the phenomenological philosophy (Husserl) – an example that is typical for such cases where it is a matter of non-immanent essences. We see here in the historical development of the space-problem how difficult it is for us in reality prejudiced humans to discover what is truly decisive. A long mathematical development was required: the great unfolding of geometrical studies from Euclid to Riemann, the physical penetration of nature and its laws since Galileo with all of its ever renewed impetus from the empirical realm, and finally the genius of individual great minds – Newton, Gauss, Riemann, Einstein – all of this was required in order to tear us away from the external, accidental, non-essential characteristics to which we would have otherwise remained attached. Certainly, if the true standpoint is once attained, a light dawns on our reason, and it then knows and recognizes what it can understand out of its own self. Nevertheless, our reason did not have the power (although during the entire development of the problem it of course always "was nearby") to see through the problem in one stroke. This must be held over against the impatience of the philosophers, who believe that they are able adequately to describe the essence on the basis of a single act of exemplary making-present. In principle they are correct, but in human terms incorrect. (1918/1922, §18)

2 For further discussion of the Helmholtz-Lie theorem, and of Weyl's generalization, see Friedman (1999). Compare also Poincaré's use of the Helmholtz-Lie theorem, discussed in Chapter 4 (this volume).

For Weyl, then, the essential structure of space can in no way be discerned "on the basis of a single act of exemplary making-present [*auf Grund eines einzigen Aktes emplarischer Vergegenwärtigung*]." On the contrary, only a lengthy historical evolution of mathematical, philosophical, and physical ideas can reveal this essential structure to us.

Why then does Weyl, like Carnap, say that his conception of the a priori "essence" of space is based on the Husserlian procedure of *Wesensanalyse*? To understand this, we must distinguish two very different ways in which space and geometry may be subject to phenomenological *Wesensanalyse*. One the one hand, geometry – taken merely as a factually given science – may be used to exemplify and motivate the ideas of *Wesenserschauung* and *Wesensanalyse* in the first place. Here we simply take it for granted that a priori geometrical knowledge is based on an intuitive grasp of immediately presented spatial forms (forms that can be intuitively given in fantasy or imagination just as well as in actual spatial perception), and we then use this idea to motivate the possibility of analogous *Wesenserschauung* in other realms – in the realm of phenomenological "pure consciousness" in particular. This use of geometry is particularly evident in Husserl (1913/1931, §70), where Husserl illustrates the idea that "immediately intuitive grasp of essences . . . can be achieved on the basis of the *mere making-present* of exemplary particulars [*auf Grund bloßer Vergegenwärtigen von exemplarischen Einzelheiten*]" precisely by the example of geometry. It is clearly this way of thinking about geometrical *Wesensanalyse* that inspires Carnap's conception in his dissertation; and it is this way, too, that appears to be the target of Weyl's complaint regarding the "impatience of the philosophers" quoted earlier.

On the other hand, however, space and geometry may be subject to phenomenological *Wesensanalyse* in a more subtle and, as it were, more constructive fashion. When we make the knowing subject or phenomenological "pure consciousness" itself into an object of *Wesenserschauung* and *Wesensanalyse*, we eventually arrive at the point where we consider the phenomenological constitution of the cognitive spatial structure – that is, the phenomenological constitution of *physical* space – in which and through which the knowing subject "looks out" at the physical world. And, as Husserl himself makes clear (1913/1931, §§136–53), the phenomenological constitution of space in this sense (i.e., of physical space) cannot be achieved on the basis of a single intuitive act of *Vergegenwärtigung*. On the contrary, since we are here dealing with the constitution of *reality*, we are necessarily involved with a never-ending approximation to a limiting idea [*Grenzidee*] in the sense of Kant. Essential knowledge of the nature of space in this sense – that is, of the physical space encountered at a definite stage of "transcendental constitution" in the "pure consciousness" of the phenomenological subject – can and indeed must be the outcome of a lengthy constructive process rather than an immediately intuitive act.

This latter kind of approach to a phenomenological *Wesensanalyse* of space – an approach only suggested in Husserl (1913/1931) – was in fact developed in great detail by Husserl's student Oskar Becker (1923). Becker's idea is to remove the "a priori contingency" that appears to attach to Euclidean geometry considered merely as a factually given "*material* eidetic science" by actually deriving or constructing Euclidean space on the basis of a phenomenological *Wesensanalyse* of the knowing subject. For, if we consider Euclidean geometry merely as a factually given "eidetic science," then, since geometry so understood is not a "*formal* eidetic science" in Husserl's sense (that is, not a branch of pure formal logic), it appears utterly "accidental" and thus "contingent" that Euclidean geometry – as opposed to some other possible geometry – actually describes the essential structure of space. Becker then attempts to remove this "a priori contingency" by constituting specifically Euclidean geometry phenomenologically on the basis of group-theoretic considerations – in an argument that goes roughly as follows: The phenomenological subject is located in a space, in which and through which it perceives the surrounding physical world. The subject must be able to move freely through this space; therefore, by the Helmholtz-Lie theorem, the space must have constant curvature. Moreover, it must be possible to distinguish rotations (by which the subject changes its orientation without changing its position) and translations (by which the subject merely changes its position); therefore, the group of rigid motions must possess a distinguished subgroup of translations, and hence the space must be Euclidean.

Now this phenomenological analysis of Becker's was of course published after Weyl's own group-theoretic work on the "space problem." Nevertheless, it appears likely that Weyl and Becker communicated about these questions much earlier, through their common participation in the phenomenological circle around Husserl in Göttingen; and, in any case, the two thinkers that Becker thanks in his Introduction are precisely Husserl and Weyl. Moreover, it seems clear that Weyl conceives his own analysis of the space problem as a generalization and refinement of Becker's – a generalization appropriate to the variably curved, non-Euclidean space(-time) of general relativity.[3]

3 Becker (1923) also considers the application of non-Euclidean, variably curved spaces in general relativity. According to Becker, however, this application must be conceived instrumentalistically with respect to the uniquely real (Euclidean) space of experience – as a device for simplifying the presentation of physical laws. For Weyl, by contrast, only the variably curved space(-time) of general relativity is the uniquely real or actual space(-time).

For Weyl, too, the phenomenological subject is located in a space, in which and through which it perceives the surrounding physical world. But Weyl does not assume that free motion is possible and thus that this space must have constant curvature. Instead, Weyl begins with the idea of an infinitesimal rotation group at every point – a group that is assumed to be the same (isomorphic) at every point, but is otherwise so far undetermined. We then fix this infinitesimal rotation group as the Euclidean-Pythagorean group by postulating that the thereby induced metric – whatever it is – must determine an associated affine connection *uniquely*. From a phenomenological point of view, then, the subject is located at a "here and now" – about which it must be able to change its orientation. By then postulating that infinitesimal translation (but not necessarily free motion) from this "here and now" is thereby uniquely determined, we guarantee that the space of our subject is *infinitesimally* Euclidean (but not necessarily either locally or globally Euclidean).

In Weyl's later discussion of space and geometry in (1927/1949), the connection between his own analysis of the space problem and the idea of phenomenological constitution is made fully explicit. In a section referring to Husserl (1913/1931) and Becker (1923) – and also, interestingly enough, to Carnap (1922) – Weyl writes:

> A way for understanding the Phythagorean nature of the metric expressed in the Euclidean rotation group precisely on the basis of the separation of a priori and a posteriori has been given by the author: Only in the case of this group does the intrinsically accidental quantitative distribution of the metric field uniquely determine in all circumstances (however it may have been formed in the context of its a priori fixed nature) the infinitesimal parallel displacement: the non-rotational progression from a point into the world. This assertion involves a deep mathematical theorem of group theory that I have proved. I believe that this solution of the space-problem plays the same role in the context of the Riemann-Einstein theory that the Helmholtz-Lie solution (section 14) plays for rigid Euclidean space. Perhaps the postulate of the unique determination of "straight-progression" can be also justified from the requirements of the phenomenological constitution of space; Becker would still like to ground the significance of the Euclidean rotation group for intuitive space on Helmholtz's postulate of free mobility. (1927/1949, §18)

This passage strongly confirms the idea that Weyl conceives his group-theoretic analysis of the space problem as a generalization and refinement of Becker's – an analysis that thereby finds its proper philosophical home within the phenomenological theory of the constitution of (physical) space.

III

Both Carnap and Weyl thus react to the new situation created by the general theory of relativity, not by adopting a straightforwardly empiricist conception of the foundations of geometry, but rather by generalizing the Kantian notion of the synthetic a priori to the *infinitesimally* Euclidean character of space. Moreover, both Carnap and Weyl base their generalization of the synthetic a priori on the Husserlian conception of *Wesensanalyse.* The two thinkers then diverge, however, on how such spatial *Wesensanalyse* is to be understood. Carnap takes (infinitesimally Euclidean) geometry, in Husserlian terms, as simply a factually given "(material) eidetic science" whose essence is revealed to us in immediately intuitive, quasi-perceptual acts of *Vergegenwärtigung* or *Wesenserschauung.* Weyl, on the other hand, explicitly rejects this kind of picture and instead views (infinitesimally Euclidean) geometry as the outcome of a complicated "transcendental constitution" by the phenomenological subject: our synthetic a priori geometrical knowledge is in no way immediately given in mere intuition but rather expresses a phenomenologically based group-theoretic construction requiring all the resources of higher mathematics.[4]

This last idea ultimately leads Weyl to distance himself even further from an immediately intuitive conception of geometry – and thus to distance himself even further from both Carnap (1922) and from Husserl (1913/1931). For Weyl comes more and more to defend the view that, although our mathematical-physical understanding of space must indeed *begin* with immediately intuitive acts (by which the subject locates and orients itself in the "here and now"), the necessary *outcome* of this procedure is a wholly nonintuitive, purely conceptual or "symbolic" construction by which we represent the physical-spatial world by abstract mathematical symbols having no intuitive content. And Weyl comes increasingly to emphasize that it is only through such purely symbolic construction, in fact, that we can obtain truly *objective* knowledge of the physical world. This becomes especially clear in a passage from Weyl (1927/1949), where, after mentioning Kant's view of space and time as mere forms of our intuition, Weyl continues:

> Intuitive space and intuitive time can therefore not serve as medium in which physics constructs the external world; [we need] rather a four-dimensional

4 Thus, the philosophical passage from the Introduction to Weyl (1918/1922) noted above puts most stress on the idea of physical reality as a limiting idea [*Grenzidee*] – in complete agreement with the conception of physical space sketched by Husserl (1913/1931, §§136–53). These sections from Husserl (1913/1931) – along with Becker (1923) – are in turn cited approvingly by Weyl (1927/1949, §18).

> continuum in the abstract-arithmetical sense. As colors for Huygens were "in reality" vibrations in the ether, so they now appear only as mathematical functional distributions of a periodic character, whereby four independent variables enter into the functions as representatives of the space-time medium via coordinates. What remains is thus finally a *symbolic construction* in precisely the sense Hilbert carries through in mathematics.
>
> The construction of this objective world, which is only presentable in symbols, from what is immediately given to me in intuition is completed in various *levels*, whereby the progress from level to level is determined by the condition that what is present at one level always reveals itself as an appearance of a higher reality – the reality of the next level. (1927/1949, §17)

This conception of a stepwise *objectification* of intuitive experience through ever more abstract, purely conceptual construction has much more in common with the neo-Kantianism of Cassirer – as expressed, for example, by Cassirer (1921/1923) – than with either Carnap (1922) or Husserlian phenomenology.

It is against this backdrop, moreover, that we should understand the striking disagreement between Weyl and Carnap over the conventionality of physical metrical structure.[5] For Weyl himself entirely rejects the idea that physical metrical structure is conventionally stipulated, and instead sees in the general theory of relativity the culmination of Riemann's suggestive remarks according to which the metric of physical space is *empirically* determined by "binding forces" acting on the underlying metrical continuum.[6] This disagreement between Weyl and Carnap should, I think, be understood in the following way. On Weyl's account, space in relativity theory has decisively transcended our spatial intuition. It is rather a purely abstract, purely conceptual structure whose "essence" or "nature" is expressed by the circumstance that the tangent space at each point of the mathematical manifold is Euclidean. It then makes perfectly good sense, for Weyl, to assert that the nonessential, entirely contingent mutual orientation of these tangent spaces is empirically determined by Einstein's field equation; for it is Einstein's theory alone that provides us with an adequate *symbolic*

5 This disagreement can be pinpointed very precisely: Carnap (1922, pp. 56–9) uses a particular form of the Schwarzschild metric due to L. Flamm to generate an alternative Euclidean description of the gravitational field involving compensating contractions in our measuring rods – thereby defending conventionalism. Weyl (1918/1922 [5th ed.], §33) uses exactly the same form of the Schwarzschild metric to argue that such alternative Euclidean descriptions, although mathematically perfectly possible, are conceptually arbitrary and therefore definitely inferior to the relativistic description.

6 Compare Weyl's well-known sixth and final *Erläuterung* to the final section of Riemann (1919).

construction of objective physical reality. On Carnap's account, by contrast, the a priori infinitesimally Euclidean nature of space is conceived of as a direct reflection of the necessary structure of our spatial intuition. And it follows, for Carnap, that specifically spatial structure (as opposed to purely formal logical-mathematical structure) can only be *intuitively* spatial structure. That this intuitively spatial structure then necessarily lacks what Weyl calls a determinate orientation of the purely infinitesimal metrics can only mean, for Carnap, that no such orientation inheres in physically spatial reality at all – in other words, that the full metrical structure of physical space can only be conventional.

Now Carnap, in his later writings, soon leaves the intuitive space of his dissertation completely behind. Indeed, Carnap (1928a/1967) adopts a stepwise constitution of objectivity via purely conceptual or logical means that is parallel to both the neo-Kantianism of Cassirer and the symbolic construction of Weyl. Physical space, in particular, becomes a purely abstract mathematical object (the set of quadrupes of real numbers $\mathbf{R}^4$) which is distinguished from other isomorphic relational structures only by the circumstance that our previously constituted epistemic subject is perceptually embedded within it – at a definite "point of view," as it were. Indeed, Carnap pushes such logical-conceptual objectification far beyond anything envisioned by Weyl; for Carnap (1928a/1967) explicitly asserts that the objectivity of science requires that *all* purely intuitive or ostensive elements must be completely and definitively expunged.[7] Accordingly, Carnap there adopts a strategy of "purely structural definite descriptions" that aims to individuate all objects of science solely on the basis of their formal or structural properties within the logic of *Principia Mathematica*; and it is in this way, finally, that Carnap ultimately breaks decisively with both neo-Kantianism and Husserlian phenomenology. But this is a topic I must leave for Part Two.

7 For Weyl, by contrast, the *origin* of our symbolic construction of the objective world in immediately intuitive experience can never be completely overcome, for setting up a coordinate system in the first place necessarily requires an *ostensive* reference to the "here and now." See, for example, the Introduction to Weyl (1918/1922): "But this objectification through the exclusion of the I and its immediate life of intuition does not succeed without remainder; the coordinate system, which can only be indicated by an individual action (and only approximately), remains as the necessary residue of this I-annihilation."

3

GEOMETRY, CONVENTION, AND THE RELATIVIZED A PRIORI: REICHENBACH, SCHLICK, AND CARNAP

Kant's analysis of scientific knowledge – as articulated especially in his *Metaphysical Foundations of Natural Science* of 1786 – is based on a sharp distinction between "pure" and "empirical" parts. The pure part of scientific knowledge consists of physical geometry (which, for Kant, is of course necessarily Euclidean geometry), more generally, the totality of applied mathematics presupposed by Newtonian physics (viz., classical analysis), Galilean kinematics (the classical velocity addition law), and the Newtonian laws of motion. In short, the entire spatiotemporal framework of Newtonian physics – what we now call the structure of Newtonian space-time – belongs to the pure part of natural science. The empirical part then consists of specific laws of nature formulated within this antecedently presupposed framework: for example, and especially, the law of universal gravitation and, more generally, the various specific force laws that can be formulated in the context of the Newtonian laws of motion.[1]

Kant holds that the pure part of scientific knowledge consists entirely of *synthetic a priori judgments*. It does not represent merely conceptual knowledge but rather results from applying the conceptual faculty of pure understanding to the distinct sensible faculty of pure intuition. (This is what Kant calls the "schematism" of the pure concepts of the understanding.) The synthetic a priori judgments belonging to the pure part of scientific knowledge then represent the *conditions of possibility* of the empirical part: the former must be in place before the latter have well-defined meaning and truth value ("relation to an object") in the first place. More generally,

1 Again, for details of Kant's philosophy of science, see Friedman (1992a,b).

the pure part of scientific knowledge represents the conditions of possibility in this sense of all empirical judgments, and thus the conditions of the possibility of experience. And it is in this way that such synthetic a priori judgments are *constitutive* of the objects of experience.

The logical positivists or logical empiricists – for present purposes, Reichenbach, Schlick, and Carnap – begin their philosophizing by emphatically rejecting this Kantian analysis of scientific knowledge and, in particular, the idea of synthetic a priori judgments. Here they are reacting to nineteenth-century work on the foundations of geometry by Gauss, Riemann, Helmholtz, Lie, Klein, and Hilbert – a development that, for the logical empiricists, culminates in Einstein's theory of relativity. In particular, Hilbert's logically rigorous axiomatization of Euclidean geometry decisively undercuts Kant's conception of the necessary role of intuition in pure mathematics, and the development of non-Euclidean geometries together with their physical application by Einstein decisively undercuts Kant's conception of applied mathematics. Moreover, relativity theory also radically revises both Galilean kinematics (viz., absolute simultaneity) and the Newtonian laws of motion. It follows that none of the principles thought to be such can be synthetic a priori in Kant's sense – indeed, they are no longer held to be even correct.

Nevertheless – and this point is not sufficiently appreciated, I think – the logical empiricists do not react to these revolutionary mathematical-physical developments by embracing an empiricist conception of geometry and, more generally, of the spatiotemporal framework of physical theory. On the contrary, they also emphatically reject the kind of conception traditionally imputed to Gauss, who is reported to have attempted to determine the curvature of physical space by measuring the angles of a terrestrial triangle determined by three mountaintops. In self-conscious opposition to such a straightforwardly empiricist conception, the logical empiricists maintain that, although Kant was wrong to think that Euclidean geometry is synthetic a priori and we can in fact use non-Euclidean geometry instead in physical theory, the question whether space is Euclidean or non-Euclidean is nonetheless not a straightforwardly empirical question. Indeed, the logical positivists here agree with Kant in rather maintaining a sharp distinction between the underlying spatiotemporal framework of physical theory, on the one hand, and the empirical laws then formulated within this framework, on the other. Their view of geometry and scientific knowledge is therefore neither strictly Kantian nor strictly empiricist.

A particularly striking version of this radically new conception of scientific knowledge is formulated by Reichenbach in his first published book

(1920/1965). According to Reichenbach, within the context of any particular given scientific theory there is a sharp and fundamental distinction between two essentially different types of principles: *axioms of coordination* and *axioms of connection.* Axioms of connection are empirical laws in the usual sense involving terms and concepts that are already sufficiently well defined. Axioms of coordination, on the other hand, are nonempirical principles that must be laid down antecedently to ensure such empirical well-definedness in the first place. Thus, for example, Gauss's proposed "experiment" to determine the geometry of physical space *presupposes* the notion of "straight line," which notion is simply not well defined independently of the geometrical and optical principles supposedly being tested. The inadequacy of such an attempt thus makes it clear that at least some geometrical principles must be laid down antecedently as axioms of coordination before any empirical determination of space even makes sense.

Accordingly, Reichenbach maintains that axioms of coordination, which paradigmatically include principles of physical geometry, are "constitutive of the concept of the object of knowledge" and that they are thus a priori in part of the Kantian sense of this term. For the Kantian notion of the a priori included two distinghuishable meanings: In the first, "a priori" means necessarily and unrevisably true, but in the second it means only constitutive in the above sense. What the development of modern geometry and relativity theory really shows is that these two meanings must be separated. Physical geometry is indeed revisable and can evolve and change with the progress of science. No geometry is necessary and true for all time. Nevertheless, at a given time and in the context of given scientific theory, the axioms of coordination – at that time and relative to that theory – are still quite distinct from the axioms of connection. The former are constitutively a priori in that, first, they are not themselves subject to straightforward empirical confirmation or disconfirmation by measuring parameters and instantiating laws, and second, they first make possible the confirmation and disconfirmation of empirical laws properly so-called (viz., the axioms of connection).

Thus, for example, in the context of Newtonian physics, Euclidean spatial geometry, Galilean kinematics, and, more generally, the structure of Newtonian space-time all count as axioms of coordination and are thus a priori in the constitutive sense *relative* to this theory. Axioms of connection include particular force laws such as the law of gravitation, whose empirical testing then presupposes that the structure of Newtonian space-time is already in place. Kant's analysis is therefore correct for Newtonian physics as an historically given theory. In special relativity, however, we change – under pressure of new empirical findings – precisely the background space-time structure.

Yet this does not mean that all principles of our new theory now count indifferently as axioms of connection. On the contrary, relative to this theory, Euclidean spatial geometry, Lorentzian kinematics, and, more generally, the structure of Minkowski space-time now count as axioms of coordination and are constitutively a priori. Straightforwardly empirical principles (axioms of connection) now include theories of particular forces and fields in Minkowski space-time, such as Maxwell's equations. In general relativity, finally, we revise our background space-time framework once again. Now, only the infinitesimally Lorentzian manifold structure – space-time topology sufficient to admit some or another (semi-)Riemannian metric – is constitutively a priori: the particular (semi-)Riemannian metrical structure realized within this framework then is determined empirically from the distribution of mass-energy, and thus the specific principles of metrical geometry now count as axioms of connection.

Reichenbach concludes that, although traditional Kantianism is thus certainly in error, traditional empiricism is equally mistaken:

> this view is distinct from an empiricist philosophy that believes it can characterize all scientific statements indifferently by the notion "derived from experience." Such an empiricist philosophy has not noticed the great difference existing between specific physical laws and the principles of coordination and is not aware of the fact that the latter have a completely different status from the former for the *logical construction* of knowledge. The doctrine of the a priori has been transformed into the theory that the logical construction of knowledge is determined by a special class of principles, and that this logical function singles out this class, the significance of which has nothing to do with the manner of its discovery and the duration of its validity. (1920/1965, pp. 93–4)

Traditional empiricism is in error precisely in failing to recognize the purely constitutive, *relativized* a priori.

Reichenbach's book contains several references to Schlick (1918/1985), and at one point (footnote 27) Schlick is explicitly criticized for, among other things, disputing "the correct part of Kant's theory, namely, the constitutive significance of the coordinating principles" (1920/1965, p. 116). Schlick wrote to Reichenbach replying to this criticism in a letter of November 26, 1920, and Reichenbach wrote back on November 29.[2] Schlick (1921/1978) then briefly alludes to this correspondence, along with

2 ASP (Archives for Scientific Philosophy), University of Pittsburgh Libraries. References are to file folder numbers. HR 015-63-22, HR 015-63-20. This exchange is usefully discussed by Coffa (1991, pp. 201–4).

Reichenbach's book, and Reichenbach (1922/1978), in turn, also alludes to the correspondence.[3]

Schlick replies in his initial letter that he entirely accepts the distinction between constitutive principles and empirical laws properly so-called. Indeed, it is only because he finds this distinction so obvious that he may have neglected sufficiently to emphasize its importance in Schlick (1918/1985). Nevertheless, Schlick prefers neither to characterize constitutive principles as a priori nor to conceive the distinction in question as in any way Kantian. For we have explicitly rejected the *synthetic* a priori along with the idea that constitutive principles possess necessary validity, and Schlick holds (quite reasonably) that these features are essential to Kant's conception of the a priori. Instead, Schlick argues, we should no longer characterize constitutive principles – for example, and especially, the principles of geometery – as a priori at all: we should rather characterize them as *conventions* in the sense of Henri Poincaré. For Poincaré had considerably earlier defended a conception of physical geometry that is neither Kantian nor empiricist – especially, of course, in Poincaré (1902/1905). Schlick himself had earlier endorsed Poincaré's conception – and extended it to the spatio*temporal* framework of physical theory – in Schlick (1915/1978) and (1917/1978). Now, in his letter, Schlick chides Reichenbach for neglecting Poincaré (Reichenbach mentions Poincaré only once, and critically, in footnote 1 to his book) and argues that the new conception of constitutive, but not synthetic, a priori principles should rest on the ideas of Poincaré rather than those of Kant. Accordingly, such constitutive principles should be characterized as conventions rather than as a priori.

Now it may seem that this dispute is merely terminological. After all, both agree in rejecting the Kantian synthetic a priori and in accepting the distinction between constitutive principles and empirical laws properly so-called, and these appear to be the only substantive questions at issue. Moreover, both Reichenbach and Schlick came to view their earlier disagreement as entirely terminological. This view of the matter is explicitly expressed in Reichenbach's letter to Schlick, in Schlick's (1921/1978) remarks, and in Reichenbach's (1922/1978) remarks. In the latter paper, Reichenbach classes himself with Schlick and Poincaré (and Einstein) as a representative of the "relativistic conception" of physical geometry, which conception

3 Schlick (1921/1978) is primarily a review of Cassirer (1921/1923). Cassirer (1921/1923, p. 460) read Reichenbach's book in manuscript, and Reichenbach (1920/1965, pp. 114–15) read Cassirer's book while his own was in press. Each expresses general approval of the other's project.

is opposed to both the "neo-Kantian" conception represented by Cassirer and the "empiricist" or "positivist" conception represented by Mach and Petzoldt – in agreement here, therefore, with the threefold conception of possible positions articulated by Schlick (1915/1978). There is no doubt, then, that Schlick's letter to Reichenbach was effective indeed, and Reichenbach adopts Poincaré's terminology of "convention" from 1922 onward – notably, in Reichenbach (1924/1969 and 1928/1958).

I want to argue, however, that the merely terminological diagnosis of their dispute adopted by both Reichenbach and Schlick is, in fact, entirely incorrect. I believe, on the contrary, that there is a most significant issue underlying their early disagreement, and that the terminological dispute obscures a fundamental difference in how the two – in 1920 – both conceive of and argue for the distinctively nonempirical status of constitutive principles such as the principles of physical geometry. Moreover, when Reichenbach – overhastily, I believe – acquiesces in the Schlick-Poincaré terminology, he also buys into Schlick's underlying conception; and the result is that the most important element in his own earlier conception of the relativized a priori is actually lost.

Schlick's conception is derived both from Poincaré and from Hilbert. From Hilbert, Schlick adopts the idea that the axioms of geometry "implicitly define" the primitive terms of that science. And this explains why the axioms of geometry are both nonempirical and conventional: alternative systems of geometry – Euclidean or non-Euclidean – simply count as different definitions of "point," "line," "between," and so on. Hilbert's view thus accounts for the nonempirical status of pure geometry. But we need to add Poincaré's ideas to account for applied geometry. Here Schlick reasons as follows: in applying such a purely formal system of implicit definitions to our actual experience of nature, no merely empirical considerations can force us to adopt one system rather than another; rather, only experience plus the requirement of overall *simplicity* of the laws of nature yields a determinate such system. The precise sense in which a determinate geometry is not forced on us by experience is expressed particularly clearly in the following passage:

> Henri Poincaré has shown with convincing clarity (although Gauss and Helmholtz still thought otherwise), that no experience can compel us to lay down a particular geometrical system, such as Euclid's, as a basis for depicting the physical regularities of the world. Entirely different systems can actually be chosen for this purpose, though in that case we also have at the same time to adopt other laws of nature. The complexity of non-Euclidean spaces can be compensated by a complexity of the physical hypotheses, and hence one can

> arrive at an explanation of the simple behavior that natural bodies actually display in experience. The reason this choice is always possible lies in the fact (already emphasized by Kant) that it is never space itself, but always the spatial behavior of *bodies*, that can become an object of experience, perception and measurement. We are always measuring, as it were, the mere product of two factors, namely the spatial properties of bodies and their physical properties in the narrower sense, and we can assume one of these two factors as we please, so long as we merely take care that the product agrees with experience, which can then be attained by a suitable choice of the other factor. (1915/1978, pp. 168–9)[4]

Schlick then explains how the requirement of overall simplicity of the total system of natural laws can single out one system of geometry in comparison with another. Essentially the same view of the matter is found in Schlick (1917/1978 and 1918/1985, §§11, 38).

In sum, Schlick's conception of the peculiarly nonempirical status of physical geometry rests on very general logical and epistemological doctrines: on Hilbert's doctrine of implicit definitions combined with what we now call Duhemian holism – viz., the idea that theoretical systems confront experience not as isolated units but only as parts of our total theory of nature.

Reichenbach's account of physical geometry (1920/1965), on the other hand, does not depend on general logical and epistemological doctrines of this kind (although Reichenbach does accept Hilbertian implicit definitions as an account of *pure* mathematics). And this is so even though Reichenbach agrees with Poincaré and Schlick that the criterion for constitutive principles is no longer necessity but rather arbitrariness:

> It is therefore not possible, as Kant believed, to single out in the concept of object a component that reason regards as necessary. It is experience that decides which elements are necessary. The idea that the concept of object has its origin in reason can manifest itself only in the fact that this concept contains elements for which *no* selection is prescribed, that is, elements that are independent of the nature of reality. The arbitrariness of these elements shows that they owe their occurrence in the concept of knowledge altogether to reason. *The contribution of reason is not expressed by the fact that the system of coordination contains unchanging elements, but in the fact that arbitrary elements occur in the system.* This interpretation represents an essential modification compared to Kant's conception of the contribution of reason. The theory of relativity has given an adequate presentation of this modification. (1920/1965, pp. 88–9)

4 This is not to say, however, that Schlick's reading of Poincaré here is accurate. For discussion, see Chapter 4 (this volume).

As this passage suggests, the kind of arbitrariness Reichenbach has in mind is quite theory specific and depends, in fact, on special features of the theory of relativity.

The idea is as follows. Each of the theories in question (Newtonian physics, special relativity, general relativity) is associated with an *invariance group of transformations* that presents us with a range of possible descriptions of nature – a range of admissible reference frames or coordinate systems – that are equivalent according to the theory. The choice of one such system over another is therefore arbitrary, and Reichenbach's thought is that those elements left invariant by the transformations in question – those elements that, as it were, mark out the range within which choice is thus arbitrary – are precisely the constitutive elements of the theory. Thus, for example, in Newtonian physics the relevant group of transformations is the Galilean group, and so, as we saw above, the underlying structure of Newtonian space-time is constitutively a priori; particular fields defined within this structure, such as the gravitational field, the distribution of mass, and so on, do not then count as constitutive. Similarly, in special relativity the relevant group of transformations is the Lorentz group, and so, as we saw above, the underlying structure of Minkowski space-time is constitutively a priori; particular fields defined within this structure, such as the electromagnetic field, the distribution of charge, and so on, do not then count as constitutive. Finally, in general relativity the relevant group includes all one-one bidifferentiable transformations (diffeomorphisms), and so only the underlying topology and manifold structure remain constitutively a priori.

To appreciate the significance of this theory specificity, it suffices to notice Reichenbach's most important result: namely, in the context of general relativity, physical geometry (the metric of physical space) is *no longer* constitutive. In Reichenbach's terminology the principles of physical geometry themselves have been transformed from axioms of coordination into axioms of connection; for, in general relativity, the metric of physical space(-time) is now dependent on the distribution of mass-energy via Einstein's field equation. For Reichenbach, then, *within* general relativity, geometry is empirical, and, in fact, Euclidean geometry is now empirically false. It follows that Reichenbach cannot accept Poincaré's (and Schlick's) conventionalism as a general philosophical doctrine about geometry as such, independently of any specific theoretical context, and it is for precisely this reason that Reichenbach explicitly rejects conventionalism in the Introduction to his book:

> mathematicians asserted that a geometrical system was established according to conventions and represented an empty schema that did not contain any

> statements about the physical world. It was chosen on purely formal grounds and might equally well be replaced by a non-Euclidean schema.[1] In the face of these criticisms the objection of the general theory of relativity embodies a completely new idea. This theory asserts simply and clearly that the theorems of Euclidean geometry do not apply to our physical space. (1920/1965, pp. 3–4)

As the footnote makes clear, Reichenbach's target here is none other than Henri Poincaré, and this explains why he discounts Poincaré's conventionalism throughout the book.

For Schlick, by contrast, geometry remains conventional or nonempirical in the context of the general theory of relativity. Duhemian holistic considerations still apply, and all that general relativity actually shows is that the *simplest* total system of natural laws employs non-Euclidean geometry: Euclidean geometry thus remains an equally "correct" option. Moreover, Reichenbach (1922/1978) comes to agree with Schlick on this crucial point,[5] and Reichenbach then appeals to such Duhemian holistic considerations in defending what he comes to call geometrical "conventionalism" throughout the remainder of his work. Reichenbach's change in terminology therefore involves also a very significant substantive change in doctrine.

This change is substantive indeed, for what I now want to emphasize is the following simple but fundamental point: Schlick's conception does not, in fact, yield a *distinction* between the constitutive and the empirical, between the conventional and factual parts of science at all – even relative to a particular given theory. Duhemian holistic considerations imply that geometry considered in isolation has no empirical consequences but only the total system of geometry plus physics has such consequences. Using Einstein's (1921/1923) notation, a given total system $G + P$ then can be empirically equivalent to an alternative system $G' + P'$. But these considerations manifestly result in no distinction between G and P, between geometry and the rest of physics. A similar result holds for the Hilbertian doctrine of implicit definitions. For a system of implicit definitions is really nothing more than a perspicuous axiomatization of a theory, and the bare idea of such an axiomatization again fails to yield a distinction between those axioms (such as the principles of geometry) that are supposed to be constitutive or conventional and those axioms (such as ordinary laws of physics or even chemistry) that are supposed to be empirical or factual.

Hence, what actually results from Schlick's conception is not conventionalism but rather *Quinean* holism: the conventional/factual distinction

5 It is also worth noting that Carnap (1922) adopts this Schlickian diagnosis of the status of geometry within general relativity as well. See Chapter 2 (this volume).

dissolves, and our total theoretical system admits no separation in principle into empirical and nonempirical parts. It follows that Schlick's conception does not yield a coherent notion of the relativized a priori at all – even apart from all terminological quibbles over whether this residue from the Kantian notion of constitutivity should be characterized as a priori or as conventional. Only Reichenbach's quite different conception (1920/1965) yields a true relativized a priori, and so, when Reichenbach accepts Schlick's view in 1922, he in fact gives up the relativized a priori.

Yet, in Carnap's (1934c) *Logical Syntax of Language* we find a revival of the relativized a priori in something very like Reichenbach's original sense. Carnap considers syntactically specified language systems or, to use his later terminology, linguistic frameworks. He emphasizes that the forms of these systems are arbitrary, and, in particular, they can embody the rules of classical logic, intuitionistic logic, or whatever. No notion of "correctness" applies to the choice of such linguistic rules, which choice is rather a purely pragmatic affair governed by the *Principle of Tolerance.* In this sense, Carnap articulates a version of Poincaré's conventionalism that is as general as possible. Moreover, the linguistic frameworks under consideration need not be purely mathematical languages, such as Carnap's Language I and Language II; they may also include physical laws. In the latter case, we are considering versions of what Carnap calls the physical language, where, in addition to purely mathematical principles, we also formulate physical hypotheses and test them via the logical-mathematical deduction of protocol-sentences.

Carnap strongly emphasizes that such testing of physical hypotheses is necessarily subject to Duhemian holistic considerations:

> Further, it is, in general, impossible to test even a single hypothetical sentence. In the case of a single sentence of this kind, there are in general no suitable [logical]-consequences of the form of protocol- sentences; hence for the deduction of sentences having the form of protocol-sentences the remaining hypotheses must also be used. Thus *the test applies, at bottom, not to a single hypothesis but to the whole system of physics as a system of hypotheses* (Duhem, Poincaré). (1934c/1937, p. 318)

Nevertheless, Carnap does not at all believe that the distinction between conventional and factual, nonempirical and empirical is thereby dissolved. How is this possible? How, that is, is Carnap essentially different from Quine?

Within any such physical language there is, for Carnap, a sharp and fundamental distinction between logical and physical principles, between what Carnap calls L-rules and P-rules, *analytic* sentences and *synthetic* sentences. This does not mean that L-rules or analytic sentences cannot be revised. On the contrary, they are just as revisable, subject to pragmatic and Duhemian holistic considerations, as are P-rules or synthetic sentences:

> No rule of the physical language is definitive; all rules are laid down with the reservation that they may be altered as soon as it seems expedient to do so. This applies not only to the P-rules but also to the L-rules, including those of mathematics. In this respect, there are only differences in degree; certain rules are more difficult to renounce than others. (1934c/1937, p. 318)

Yet there is, in another respect, an essential difference in kind between the two types of theoretical revision. Changing the L-rules involves changing the language and therefore the meanings of the terms of the language, whereas changing only the P-rules involves no such change of language but only a revision of synthetic or empirical sentences formulated within a given (and therefore fixed) language.

I suggest that Carnap's L-rules or analytic sentences can be profitably viewed as a precise explication of Reichenbach's notion of the constitutive or relativized a priori. And, in this connection, it is especially interesting to note that Carnap's L-rules include not only pure mathematics but also, in the context of some linguistic frameworks, principles of physical geometry. More precisely, it is Carnap's view that, in the context of physical geometries of constant curvature, the term for the metric constitutes a logical rather than a descriptive expression and thus the principles of metrical geometry are L-rules or analytic sentences; in the case of the variable, mass-energy–dependent curvature of general relativity, however, the term for the metric constitutes a descriptive expression and the principles of metrical geometry are P-rules or synthetic sentences.[6] Thus, Carnap's analysis of the transition from Newtonian physics and special relativity to general relativity here agrees precisely with Reichenbach's analysis of 1920: geometry has itself undergone a transition from a nonempirical and constitutive status to an empirical and thus nonconstitutive status – indeed, Euclidean geometry is now empirically, that is, synthetically, false.[7]

6 See Carnap (1934c/1937, pp. 178–9). Intuitively, the descriptive expressions are such that all sentences built up from them alone are *determinate* relative to the rules of the framework in question; L-rules or synthetic sentences then are those theorems of the framework that are invariant under substitution of descriptive expressions. The metric in general relativity counts as descriptive because sentences ascribing particular values of curvature to particular space-time points are not fixed by general laws of the theory but only by the latter plus a specific distribution of mass-energy.

7 And it is again worth noting that this contrasts sharply with Carnap's 1922 analysis of general relativity based on the conception of Schlick and Poincaré: see footnote 5, this chapter.

Carnap's account, like Reichenbach's, thus yields a theory specific (language specific) distinction between two intrinsically different types of principles and therefore a true relativized a priori. Unlike Reichenbach's account, however, which is applicable only to the classical and relativistic space-time theories through the machinery of invariance groups and coordinate systems, Carnap's notion of L-rules or analytic sentences is applicable to arbitrary formally specified languages or theories. Moreover, Carnap, again unlike Reichenbach, has a general explanation of what constitutivity amounts to: viz., constitutive principles are constitutive of the language in question and thus of the meanings of the terms of the language. In Reichenbach's original account, by contrast, no real link between the intuitive notion of constitutivity and the machinery of invariance groups and coordinate systems has actually been forged. Carnap's account, although it, too, is in some sense purely formal or purely logical, is also far superior to an attempted explanation of constitutivity via the Hilbertian notion of implicit definitions. For Carnap's analysis depends precisely on a distinction or differentiation of the axioms of a given theory into two essentially different types – L-rules and P-rules – and thus we do not get the unwanted result that all axioms indifferently are constitutive or definitive of the language in question.

Finally, as we have seen, Carnap is able to accept the basic idea lying behind Duhemian holism without dissolving the desired distinction between conventional and factual, constitutive and empirical. And this shows, it seems to me, that Duhemian holism and the general revisability of all theoretical principles are together quite insufficient to yield the collapse of the empirical/nonempirical distinction and thus *Quinean* holism. For Carnap's *Logical Syntax* program incorporates both of these ideas and also a sharp distinction between L-rules and P-rules, analytic sentences and synthetic sentences. Indeed, as we have seen, Reichenbach in 1920 had already articulated a conception according to which all theoretical principles are in fact revisable while a sharp distinction between the a priori and the empirical is nonetheless maintained: this is in fact the entire point of the *relativized* a priori. To obtain Quinean holism from this starting point we must therefore finally exhibit the incoherence of Carnap's *Logical Syntax* program, and only this, I suggest, demonstrates the ultimate failure of the logical positivists' version of the relativized a priori.

4

POINCARÉ'S CONVENTIONALISM AND THE LOGICAL POSITIVISTS

The great French mathematician Henri Poincaré is also well known, in philosophical circles, as the father of geometrical conventionalism. In particular, the logical positivists appealed especially to Poincaré in articulating and defending their own conception of the conventionality of geometry. As a matter of fact, the logical positivists appealed both to Poincaré and to Einstein here, for they believed that Poincaré's philosophical insight had been realized in Einstein's physical theories. They then used both – Poincaré's insight and Einstein's theories – to support and to illustrate their conventionalism. They thus viewed the combination of Poincaré's geometrical conventionalism and Einstein's theory of relativity as a single unified whole.

How, then, do the logical positivists understand Poincaré's argument? They concentrate on the example Poincaré (1902/1905) presents in the fourth chapter of *Science and Hypothesis*: the example, namely, of a world endowed with a peculiar temperature field. According to this example, we can interpret the same empirical facts in two different ways. On the one hand, we can imagine, in the given circumstances, that we live in an infinite, non-Euclidean world – in a space of constant negative curvature. On the other hand, we can equally well imagine, in the same empirical circumstances, that we live in the interior of a finite, Euclidean sphere in which there also exists a special temperature field. This field affects all bodies in the same way and thereby produces a contraction, according to which all bodies – and, in particular, our measuring rods – become continuously smaller as they approach the limiting spherical surface. (Poincaré of course obtains the law of this contraction from his own model of Bolyai-Lobachevsky space.) We thus are confronted here with a case of observational equivalence, and so, no empirical facts can force us to select either the Euclidean or the non-Euclidean description as the uniquely correct description. In this sense the choice of geometry is entirely free and therefore conventional.

Moritz Schlick, the founder of the Vienna Circle, presents just such an interpretation of Poincaré's argument in his 1915 article on the philosophical significance of the theory of relativity, which was the first article on relativity theory within the tradition of logical positivism:

> Henri Poincaré has shown with convincing clarity (although Gauss and Helmholtz still thought otherwise), that no experience can compel us to lay down a particular geometrical system, such as Euclid's, as a basis for depicting the physical regularities of the world. Entirely different systems can actually be chosen for this purpose, though in that case we also have at the same time to adopt other laws of nature. The complexity of non-Euclidean spaces can be compensated by a complexity of the physical hypotheses, and hence one can arrive at an explanation of the simple behavior that natural bodies actually display in experience. The reason this choice is always possible lies in the fact (already emphasized by Kant) that it is never space itself, but always the spatial behavior of *bodies*, that can become an object of experience, perception and measurement. We are always measuring, as it were, the mere product of two factors, namely the spatial properties of bodies and their physical properties in the narrower sense, and we can assume one of these two factors as we please, so long as we merely take care that the product agrees with experience, which can then be attained by a suitable choice of the other factor. (1915/1978, pp. 168–9)

This argument and relativity theory fit together especially well, according to Schlick, because relativity theory also is based on the idea that space and matter cannot be separated from one another.

Approximately fifty years later, we find Rudolf Carnap still presenting essentially the same argument in his *Introduction to the Philosophy of Science*:

> Suppose, Poincaré wrote, that physicists should discover that the structure of actual space deviated from Euclidean geometry. Physicists would then have to choose between two alternatives. They could either accept non-Euclidean geometry as a description of physical space, or they could preserve Euclidean geometry by adopting new laws stating that all solid bodies undergo certain contractions and expansions. As we have seen in earlier chapters, in order to measure accurately with a steel rod, we must make corrections that account for thermal expansions or contractions of the rod. In a similar way, said Poincaré, if observations suggested that space was non-Euclidean, physicists could retain Euclidean space by introducing into their theories new forces – forces that would, under specified conditions, expand or contract the solid bodies. (1966/1974, pp. 144–5)

Carnap then concludes this chapter on Poincaré's philosophy of geometry by remarking that we will see in the next two chapters on relativity theory how Poincaré's insight into the observational equivalence of Euclidean and non-Euclidean theories of space leads to a deeper understanding of the structure of space in relativity theory (1966/1974, p. 151).

In my opinion, however, this conception of the relationship between Poincaré and Einstein rests on a remarkable – and, in the end, ironical – misunderstanding of history. The first point to notice is that the logical positivists' argument from observational equivalence is in no way a good argument for the conventionality of geometry, at least as this was understood by Poincaré himself. For the argument from observational equivalence has no particular relevance to physical geometry and can be applied equally well to *any* part of our physical theory. The argument shows only that geometry considered in isolation has no empirical consequences: such consequences are only possible if we also add further hypotheses about the behavior of bodies. But this point is completely general and is today well known as the Duhem-Quine thesis: *all* individual physical hypotheses require further auxiliary hypotheses in order to generate empirical consequences.[1]

Poincaré's own conception, by contrast, involves a very special status for physical geometry. He emphasizes in the Preface to *Science and Hypothesis*, for example, that his leading idea is that hypotheses of different kinds should be carefully distinguished from one another:

> We will also see that there are various kinds of hypotheses; that some are verifiable and, when once confirmed by experiment, become truths of great fertility; that others, without being able to lead us into error, become useful to us in fixing our ideas, and that the others, finally, are hypotheses in appearance only and reduce to definitions or conventions in disguise. (1902/1905, p. xii)

Poincaré then enumerates the sciences where we are involved principally with the free activity of our own mind: arithmetic, the theory of mathematical magnitude, geometry, and the fundamental principles of mechanics. At the end of the series of sciences, however, comes something quite different, namely, experimental physics. Here we are certainly involved with more than our own free activity:

> Up to here [mechanics] nominalism triumphs, but we now arrive at the physical sciences properly speaking. Here the scene changes: we meet with hypotheses of another kind, and we recognize their great fertility. No doubt at first sight our theories appear fragile, and the history of science shows us how

1 This point is emphasized in Chapter 3 (this volume).

> ephemeral they are; but they do not entirely perish, and from each of them something remains. It is this something that it is necessary to try to discover, because it is this, and this alone, that is the true reality. (1902/1905, p. xxvi)

The fourth part of *Science and Hypothesis* explicitly considers precisely these physical sciences properly speaking. There, under the heading "Nature," Poincaré discusses what he takes to be genuinely physical theories: optics and electrodynamics, for example. Despite the obvious fact that the above-mentioned Duhemian argument applies equally well to these theories as well, Poincaré nevertheless considers them to be *non*conventional. Hence, this Duhemian argument certainly cannot – at least by itself – be Poincaré's own argument for the conventionality of geometry.

Poincaré's own argument involves two closely related ideas. The first is the already indicated idea that the sciences constitute a series or a hierarchy. This hierarchy begins with the purest a priori science – namely, arithmetic – and continues through the above-mentioned sciences to empirical or experimental physics properly speaking. In the middle of this hierarchy, and thus in a very special place, we find geometry. The second idea, however, is the most interesting and important part of Poincaré's argument. For Poincaré himself is only able to argue for the conventionality of geometry by making essential use of the Helmholtz-Lie solution to the space problem. This specifically group-theoretical conception of the essence of geometry, that is, is absolutely decisive – and thus unavoidable – in Poincaré's own argument. In what follows, we consider these two ideas more closely.

The series or hierarchy of sciences begins, as we said, with arithmetic. For Poincaré, arithmetic is not, of course, a branch of logic, for logic is a purely analytical science and thus purely tautological, whereas arithmetic is the first and foremost synthetical science – which therefore genuinely extends our knowledge. Arithmetic is synthetic because it is based on our intuitive capacity to represent the (potentially) infinite repetition of one and the same operation. And this intuition is then the ground for the characteristically mathematical procedure of reasoning, namely, mathematical induction or reasoning by recurrence. Such reasoning by recurrence comprehends as it were an infinite number of syllogisms and is precisely for this reason in no way merely analytic. For no merely analytical procedure can possibly lead us from the finite to the infinite. Nevertheless, arithmetic is wholly a priori as well: mathematical induction forces itself upon us uniquely and necessarily because it is precisely the expression of a unique power of our own mind. Therefore, arithmetic is neither an empirical science nor a conventional one.

The next lower level in the hierarchy of sciences is occupied by the theory of mathematical magnitude. Here Poincaré considers what we nowadays refer to as the system of real numbers. Poincaré, however, is not only interested in the purely formal properties of this system; on the contrary, he is interested above all in the psychological-empirical origin of our concept of this system. Specifically, he explains the origin of our concept of the system of real numbers in two steps. He first describes how the idea of the continuum arises, namely, through the repeated or iterative application of the principle of noncontradiction to just noticeable differences in Fechner's sense. But here we have only obtained the idea of an order continuum, which does not yet contain metrical or measurable magnitudes. Then, to construct the latter, we must introduce a further element, namely, an addition operation. And, according to Poincaré, the introduction of such an addition operation is almost entirely arbitrary. Of course, it must satisfy certain conditions – the conditions for a continuous, additive semigroup. Nevertheless, according to Poincaré, we are entirely free to introduce any addition operation whatsoever that satisfies the given formal conditions. Here, therefore, for the first time, we have a convention properly speaking, that is, a free stipulation.

So far we have considered only one-dimensional continua. When we attempt to apply these ideas to multidimensional continua, we reach the next level in the hierarchy of sciences, namely, the science of geometry. A multidimensional continuum becomes an object of geometry when one introduces a metric – the idea of measurability – into such a continuum. And, analogously to the case of one-dimensional continua, we achieve this through the introduction of group-theoretical operations. In this case, however, the structure of the operations in question is much more interesting from a mathematical point of view. In the case of a three-dimensional continuum, for example, instead of a continuous, additive semigroup of one dimension, we have a continuous group of free motions (in modern terminology, a *Lie group*) of six dimensions. And, in my opinion, we can achieve a deeper understanding of Poincaré's own conception of the conventionality of geometry only through a more careful consideration of precisely these group-theoretical structures.

We will come back to this question in a moment. First, however, it is necessary briefly to consider the remaining two levels in the hierarchy of sciences. The next lower level after geometry is occupied by the science of mechanics. The laws of mechanics – for example, the Newtonian laws of motion – govern the fundamental concepts of time, motion, mass, and force; and these laws, according to Poincaré, are also conventional, at least for the most part. I understand him here to be arguing that the fundamental concepts of time, motion, mass, and force have no determinate empirical meaning *independently* of the laws of mechanics. Thus, for example, the laws of motion supply us with an implicit definition of the inertial frames of reference, without which no empirically applicable concept of time or motion is possible; the concepts of mass and force are only empirically applicable on the basis of the second and third Newtonian laws of motion; and so on. The laws of mechanics do not therefore describe empirical facts governing independently given concepts. On the contrary, without these laws, we simply would have no such concepts: no mechanical concepts, that is, of time, motion, mass, and force. In this sense the laws of mechanics are also free creations of our mind, which we must first inject, as it were, into nature.

Now, however, we have finally reached the empirical laws of nature properly speaking. For we have now injected precisely enough structure into nature to extract the genuinely empirical laws from nature. We do this, for example, by discovering particular force-laws that realize the general concept of force defined by the laws of mechanics. Poincaré himself considers in this connection the Maxwell-Lorentz theory of the electromagnetic field and electrodynamic force especially, for this theory was, of course, of most interest in his time. But the point perhaps can be made even more clearly if we consider Newton's theory of universal gravitation. For Newton's *Principia* had already shown clearly how we can empirically discover the law of universal gravitation – on the *presupposition*, that is, of the Newtonian laws of motion and Euclidean geometry. Without these presuppositions, however, we certainly would not have been able to discover the law of gravitation. And the same example also shows clearly how *each* level in the hierarchy of sciences presupposes *all* of the preceding levels: we would have no laws of motion if we did not presuppose spatial geometry, no geometry if we did not presuppose the theory of mathematical magnitude, and of course no mathematics at all if we did not presuppose arithmetic.

We now return to a more detailed consideration of geometry. The metrical properties of physical space are based, as indicated above, on a Lie group of free motions; and the idea of such a group arises, according to Poincaré, from our experience of the motion of our own bodies. We thereby learn, in particular, to distinguish between changes in external objects and changes (i.e., motions) of our own bodies. Then, through an idealization, we construct a separate concept of these latter changes (motions of our own bodies), and we represent this concept by means of a mathematical group. In this sense – that is, through an idealization – the idea of such a Lie group arises from our experience. At this point, however, a remarkable mathematical theorem comes into play, namely, the Helmholtz-Lie theorem. For, according to his theorem, there are three and only three possibilities for such a group: either it can represent Euclidean geometry (i.e., it is a group of free motions of rigid bodies in a Euclidean space), or it can represent a geometry of constant negative curvature (hyperbolic or Bolyai-Lobachevsky space), or it can represent a geometry of constant positive curvature (elliptic or, as it is sometimes called, Riemannian space). What is important here, for Poincaré, is that only the idea of such a Lie group can explain the origin of geometry, and, at the same time, this idea drastically restricts the possible forms of geometry.

Poincaré, of course, believes that the choice of any one of the three groups is conventional. Whereas experience suggests to us the general idea of a Lie group, it can in no way force us to select a specific group from among the three possibilities. Analogously to the case of the theory of mathematical magnitude, we are here concerned basically with the selection of a standard measure or scale:

> This is the object of geometry: it is the study of a particular "group"; but the general concept of a group preexists in our mind, at least potentially. It imposes itself upon us – not as a form of our sensibility, but as a form of our understanding. However, from among all possible groups it is necessary to choose one that will be so to speak the *standard measure* [*étalon*] to which we relate the phenomena of nature.
>
> Our experience guides us in this choice but does not impose it upon us; it allows us recognize, not which is the truest geometry, but rather which is the most *convenient*. (1902/1905, pp. 70–1)

> In our mind the latent idea of a certain number of groups preexists: those for which Lie has supplied the theory. Which shall we choose to be a kind of standard measure by which to compare the phenomena of nature?.... Our experience has guided us by showing us which choice is best adapted to the properties of our own body. But there its role ends. (pp. 87–8)

But such a selection in this case is much more interesting from a mathematical point of view. In contrast to the case of one-dimensional continua, a selection of the relevant group-theoretical operations here determines that the resulting system has one (and only one) of the three possible mathematical structures (Euclidean, constant negative curvature, or constant positive curvature). In this sense the mathematical laws here are completely determined by the selection of a particular scale.

Poincaré's conception becomes clearer when we contrast it with Helmholtz's earlier conception of geometry. For Helmholtz, of course, also proceeds from such group-theoretical considerations – that is, from the possibility of free motion – in attempting to justify a more empiricist conception of geometry; and, for precisely this reason, Helmholtz gives the title, "On the *Facts* which Lie at the Basis of Geometry," to his main contribution here. Where, then, lies the disagreement between Helmholtz and Poincaré? We should first remind ourselves that Helmholtz had first left Bolyai-Lobachevsky geometry completely out of consideration. His original idea was that there are only two possible geometries, namely, Euclidean geometry and elliptical (or spherical) geometry. From the fact that free motion in general is possible, it follows that space must be either Euclidean or spherical. From the further fact that free motion is possible *to infinity* (so that an infinite straight line is possible), it then follows that space must be Euclidean. Now Helmholtz, of course, soon corrected this erroneous idea when he became acquainted with Bolyai-Lobachevsky geometry (through the work of Beltrami); Poincaré, by contrast, clearly recognized from the very beginning that the most important and interesting choice is that between Euclidean and Bolyai-Lobachevsky geometry.[2]

In the second place, however, Poincaré also clearly saw that the idea of the free motion of rigid bodies is itself an idealization: strictly speaking, there are in fact no rigid bodies in nature, for actual bodies are always subject to actual physical forces. It is therefore completely impossible simply to read off, as it were, geometry from the behavior of actual bodies without first formulating theories about physical forces. (In my opinion, the point of the temperature-field example is precisely to make *this* situation intuitively clear.) And it now follows that geometry cannot depend on the behavior of actual bodies. For, according to the above-described hierarchy of sciences, the determination of particular physical forces presupposes the laws of motion, and the laws of motion in turn presuppose geometry itself: one must first set up a geometry before one can establish a particular theory of physical forces. We have no other choice, therefore, but to select one or another geometry on conventional grounds, which we then can use, so to speak, as a standard measure or scale for the testing and verification of properly empirical or physical theories of force. Moreover, it is also remarkable (and we return to this point later) that relativity theory confirms Poincaré's conception more than it does Helmholtz's. For we here apply non-Euclidean

2 For a discussion of Helmholtz's views, including the important change when he learned of Bolyai-Lobachevsky geometry, see Friedman (1997). For further discussion of the relation between Helmholtz and Poincaré, see Friedman (1999). Some of Helmholtz's central works on geometry are collected by Hertz and Schlick (1921/1977).

geometry to nature, not through the mere observation of the behavior of rigid bodies, but rather through a fundamental revision of both the laws of motion and our physical theory of gravitation.

Nevertheless, relativity theory also shows that Poincaré's own conception of the role of geometry in physics is false in principle. For Poincaré's conception is based entirely, as we have seen, on an application of the Helmholtz-Lie theorem: geometry is conventional precisely because the general idea of a Lie group of free motions has three (and only three) possible geometrical realizations. Poincaré therefore presupposes throughout that the free motion of an ideal rigid body is possible and hence that space is homogeneous and isotropic: the only geometries that are possible in Poincaré's conception are the classical geometries of constant curvature. By contrast, in the general theory of relativity, we use the much more general conception of geometry articulated in Riemann's theory of manifolds (not to be confused, of course, with the very particular case of constant positive curvature, which sometimes is also called Riemannian geometry). According to the general theory of relativity, space (more precisely, the space-time continuum) is a manifold of *variable* curvature – and, in fact, a curvature that depends essentially on the distribution of matter.

Poincaré was not, of course, acquainted with the general theory of relativity. (He died in 1912.) He is nevertheless completely clear that his conception of geometry is not compatible with Riemann's theory of manifolds. And, for precisely this reason, he considers this more general theory to be purely analytical:

> If, therefore, one admits the possibility of motion, then one can invent no more than a finite (and even rather restricted) number of three dimensional geometries. However, this result appears to be contradicted by Riemann; for this scientist constructs an infinity of different geometries, and that to which his name is ordinarily given is only a special case.... This is perfectly exact, but most of these definitions [of different Riemannian metrics] are incompatible with the motion of an invariable figure – which one supposes to be possible in Lie's theorem. These Riemannian geometries, as interesting as they are in various respects, can therefore never be anything but purely analytic, and they would not be susceptible to demonstrations analogous to those of Euclid. (1902/1905, pp. 47–8)[3]

3 In the Greenstreet translation (Poincaré 1902/1905), the last sentence is incorrectly translated as: "These geometries of Riemann, so interesting on various grounds, can never be, therefore, purely analytical, and would not lend themselves to proofs analogous to those of Euclid" – thereby entirely reversing its sense (and the preceding sentence on p. 47 incorrectly has "variable figure" instead of "invariable figure").

The Riemannian theory is purely analytical because it is not based on group-theoretical operations and therefore not on the possibility of repeating a given operation indefinitely:

> *Space is homogeneous and isotropic.* One may say that a motion that is produced once can be repeated a second time, a third time, and so on, without changing its properties. In the first chapter, where we studied the nature of mathematical reasoning, we have seen the importance that one should attribute to the possibility of repeating indefinitely the same operation. It is in virtue of this repetition that mathematical reasoning acquires its force; it is thanks to the law of homogeneity that it applies to the facts of geometry. (1902/1905, p. 64)

Poincaré's conception is therefore entirely self-consistent, for the Riemannian manifolds of variable curvature conflict with his explanation of the fact that geometry is a properly synthetic science.

The general theory of relativity, however, contradicts Poincaré's conception in a second, and even more fundamental, respect. This theory describes the motion of a body in a gravitational field as a geodesic (straightest possible curve) in a four-dimensional manifold, that is, as a geodesic in a space-time continuum possessing a variable curvature depending explicitly on the distribution of matter. And this completely new formulation of the law of gravitation then also takes over the role previously played by the laws of motion, for the geodesics in space-time traversed by bodies in a gravitational field have here precisely the role previously played by the inertial motions. In other words, the law of gravitation takes over here the role of the law of inertia. It then follows, however, that one can no longer separate geometry from the laws of motion, and one can no longer separate the latter from the law of gravitation. On the contrary, in the general theory of relativity, geometry is simply identical to the theory of gravitation; this theory is, in turn, identical to the laws of motion or mechanics; and geometry is therefore also identical to mechanics.

In the general theory of relativity, there can therefore be no question of a hierarchy of sciences in Poincaré's sense. Poincaré presents mathematical physics as a series of sciences in which every succeeding science presupposes all preceding sciences. General mechanics is presupposed by particular force-laws and thus makes the latter possible; geometry is presupposed by general mechanics and thus makes both it and particular force-laws possible; the theory of mathematical magnitude is presupposed by geometry; and arithmetic is presupposed by the theory of mathematical magnitude. In this way, Poincaré's conception of the sciences is actually quite similar to the Kantian conception. Yet Poincaré is writing at the end of the nineteenth century and therefore cannot proceed from the idea that Euclidean geometry is the only possible geometry. In the context of the Helmholtz-Lie solution to the space problem, it then appears natural to suppose that we have a conventional choice among three (and only three) possibilities. And, precisely because geometry still appears to be the presupposition of all properly empirical sciences, this choice cannot itself be empirical. Thus, Poincaré's modernized Kantianism is particularly well adapted to the scientific situation of the late nineteenth century – such a modified Kantianism can no longer be maintained in the context of the radically new physics of the twentieth century, however.

In contrast to Poincaré, it is clear that the logical positivists, for their part, belong entirely to the twentieth century. And, in fact, Rudolf Carnap, Hans Reichenbach, and Moritz Schlick all attempted in their earliest writings philosophically to comprehend the theory of relativity. They even undertook the task of fundamentally reforming philosophy itself through precisely this attempt to comprehend Einstein's physical theories. Thus, for example, from the very beginning the logical positivists explicitly asserted that Einstein's new theories are completely incompatible with the Kantian conception of the synthetic a priori, so that this philosophical conception now is simply untenable. They also clearly recognized that Helmholtz's geometrical empiricism is untenable as well. For, in the general theory of relativity, we construct a non-Euclidean description of nature (as emphasized earlier), not by simply observing the behavior of rigid measuring rods, but rather by fundamentally revising both general mechanics and our theory of gravitational force. The logical positivists therefore sought for an intermediate position, as it were, lying *between* traditional Kantianism and traditional empiricism. And it seemed to them that precisely such an intermediate position was to be found in Poincaré's conception of convention.

We have seen, however, that Poincaré's own argument for geometrical conventionalism actually fails in the context of the general theory of relativity: neither his conception of a hierarchy of sciences nor his penetrating and insightful application of the Helmholtz-Lie theorem make sense in this new conceptual framework. The general theory of relativity essentially employs a geometry of variable curvature and also effects a holistic unification of previously separated sciences. For the logical positivists, there was therefore no alternative but simply to ignore the characteristic elements of Poincaré's own argument and to concentrate instead solely on the example of the peculiar temperature field. In the absence of Poincaré's own conception of a hierarchy of sciences, however, it is clear that this example by itself can have no *particular* relevance to geometry. On the contrary, we thereby obtain (as emphasized at the very beginning) only a completely general holism, according to which *every* individual scientific hypothesis has empirical consequences only in connection with further auxiliary hypotheses. In other words, we thereby obtain only what is nowadays referred to as Duhemian or Duhem-Quine holism. And Quine himself, as is well known, uses this Duhemian holism precisely to attack the conventionalism of the logical positivists: according to Quine there is, of course, no longer a difference in principle between facts on the one side and conventions on the other. It is therefore extremely problematic, at best, to base the thesis of the conventionality of geometry on Duhemian holism. As we have seen, what is most ironical here is the circumstance that just this holistic collapse of the conventional/factual distinction was already prefigured in the earlier encounter between Poincaré's geometrical conventionalism and the general theory of relativity.

It is therefore noteworthy that there was one logical positivist who, at least once in his life, correctly and explicitly recognized the incompatibility of Poincaré's conventionalism with the general theory of relativity. This was Hans Reichenbach, in his (1920/1965) *Theory of Relativity and A Priori Knowledge*:

> It was from a mathematical standpoint asserted that geometry has only to do with conventional stipulations – with an empty schema containing no statements about reality but rather chosen only as the form of the latter, and which can with equal justification be replaced by a non-Euclidean schema.[1] Against these objections, however, the claim of the general theory of relativity presents a completely new idea. This theory makes the equally simple and clear assertion that the propositions of Euclidean geometry are just *false*.
>
> [1]Poincaré has represented this view. Cf. [*Science and Hypothesis*, Chapter III]. It is significant that for his proof of equivalence he excludes from the beginning *Riemannian* geometry, because it does not permit the displacement of a body without change of form. If he had guessed that precisely this geometry would be taken up by physics, he would never have been able to assert the arbitrariness of geometry. (1920/1965, pp. 3–4 & note 1, p. 109)

Unfortunately, Reichenbach was soon convinced by Schlick that Poincaré's conception could still be valid in the context of the general theory of relativity.[4] As is well known, Reichenbach then occupies himself, in his

4 This is discussed further in Chapter 3 (this volume). It does not follow, however, that general relativity is incompatible with the more general idea of constitutive principles as such

later writings, precisely with the attempt to combine relativity theory with conventionalism. That this attempt must fail is implicit in the analysis of Poincaré's conventionalism that I have presented.

Here, however, I will not pursue the story of Reichenbach's later conventionalism further. But I do want to emphasize how far the basic philosophical conception of the logical positivists deviates from that of Poincaré himself. For the empiricism of the logical positivists consists in precisely the circumstance that they completely reject the Kantian doctrine of synthetic a priori judgments. In their case the concept of convention is then a substitute for the synthetic a priori that is supposed to take over the function of the Kantian a priori in all domains of thought: they apply the concept of convention, not only to comprehend physical geometry, but also to explain pure mathematics and even logic. According to the logical positivists, all a priori sciences rest in the end on conventional stipulations – and precisely in this way is Kantianism once and for all decisively overcome.

By contrast, Poincaré himself gives a central place to the synthetic a priori. In fact, as we have seen, his conception of arithmetic is extremely close to the original Kantian conception of arithmetic. First, arithmetic is based on our intuitive capacity for representing the indefinite repetition or iteration of one and the same operation, and therefore arithmetic for Poincaré is not a merely analytic science. Second, arithmetic is also not conventional for Poincaré: mathematical induction necessarily forces itself upon us, and there are thus no alternatives here. Third, arithmetic occupies the apex or summit of a hierarchy of sciences: all other sciences – all other a priori sciences, in particular – presuppose arithmetic because all others presuppose mathematical induction or reasoning by recurrence.

Now, Poincaré's conception of geometry is also very similar to the Kantian conception of geometry. For Poincaré, as for Kant, geometry is synthetic because it is based, like arithmetic, on the possibility of indefinitely repeating particular operations, namely, group-theoretical operations constituting a Lie group of free motions. Moreover, geometry also is viewed as the presupposition of all properly empirical physical theories: neither for Poincaré nor for Kant can geometry itself be either empirically confirmed or empirically disconfirmed. The difference, of course, is that Poincaré, in contrast to Kant, is acquainted with *alternative* geometries. Poincaré is acquainted, in

(axioms of coordination in Reichenbach's sense). Although the specific metrical structure of space(-time) now appears as empirical, it can still plausibly be held (as Reichenbach held in 1920) that the topology, differentiable structure, and (semi-)Riemannian form of the metric are constitutive. Compare also the discussion of Carnap's dissertation and Weyl in Chapter 2 (this volume).

particular, with the Helmholtz-Lie theorem, according to which geometry is constrained, but by no means uniquely determined, by the idea of a Lie group of free motions. It then follows for Poincaré, because three alternative possibilities are still left open, that we have here – in this very special situation – a conventional choice or free stipulation.[5]

Poincaré's basic philosophical conception thus by no means implies a general rejection of the synthetic a priori. On the contrary, without the synthetic a priori, his argument simply makes no sense. Precisely because geometry, like arithmetic, is synthetic, but also – according to the Helmholtz-Lie theorem and in contradistinction to arithmetic – is not uniquely determined, it follows that geometry is conventional. For the logical positivists, by contrast, there can be no question of *this* kind of argument for geometrical conventionalism. Because arithmetic is no longer viewed as synthetic a priori in the Kantian sense, they, for their part, attach no particular importance to our intuitive capacity for representing the indefinite repetition of some or another operation. Moreover, because we now consider geometry first and foremost in the context of the Riemannian theory of manifolds, the Helmholtz-Lie theorem is no longer relevant in any case.[6] And, finally, we now accept the general theory of relativity (indeed, as the very paradigm of a successful physical theory); and, according to this theory, there is no longer any possibility of conceiving geometry as the presupposition of properly empirical physics. As we have seen, we are, in fact, forced by this theory to subscribe to a holistic conception of the relationship between geometry and empirical physics. Before the development of the general theory of relativity, we were, of course, free to adopt such a holistic conception if we wished – but, after this development, there simply could be no alternative.

The main point of our earlier discussion, however, is that such a holism is much too weak to support a special, nonempirical status for geometry. Holism by itself is obviously also completely unable to explain the nonempirical status of arithmetic. If the logical positivists really wish to apply the concept of convention as an explanation of the status of the a priori in general, therefore, they clearly need to add some entirely new element that goes beyond mere holism. And this, in fact, is precisely what happens: When Rudolf Carnap (1934c) then attempts to articulate a general conventionalistic conception of the a priori, holism plays only a very subsidiary

5 For further discussion of the relation between Poincaré and Kant, see Friedman (1999).

6 But see the discussion of Weyl's generalization of the Helmholtz-Lie theorem in Chapter 2, §II (this volume).

role. Instead, everything depends on the new conception of *analyticity* that he attempts to develop there.

Carnap considers purely formal languages or linguistic frameworks that can be chosen entirely arbitrarily. We can, for example, choose a language governed by the rules of classical (Frege-Russell) logic; but we can also, with equal justification, choose an entirely different type of language governed by the rules of intuitionistic logic. In fact, there can be no question here of either "justification" or "correctness," for the very concept of correctness itself only has meaning when we have antecedently specified a particular linguistic framework. Hence, the choice of one or another such framework can be based only on a convention, which we stipulate entirely freely on pragmatic grounds. What is most important, however, is the following: Relative to any particular formal language or linguistic framework, there is a sharp distinction between the logical rules or analytic sentences of the framework, and the physical rules or synthetic sentences of the framework. In particular, the former constitute the underlying logic of the framework, which first makes questions of correctness, justification, and so on possible. Our conventional choice of a language, together with the characteristic logical rules of this language, then clarifies the special epistemological (and nonempirical) status of such rules.

Carnap does not therefore represent a general holism, according to which all sentences whatsoever have precisely the same status: instead, we are given a sharp distinction between logical and physical rules – analytic and synthetic sentences. Within a framework for classical mathematical physics, for example, (classical) logic, arithmetic, and the theory of real analysis belong to the logical rules, whereas Maxwell's field equations belong to the physical rules. The former are therefore conventional in the context of this framework, whereas the latter are nonconventional and thus empirical. And what is the status of geometry here? From the present point of view, Carnap's result is especially interesting and noteworthy. Within a framework for classical mathematical physics, in which space has constant curvature, geometry also belongs to the logical (or analytic) rules. Within a framework like that of the general theory of relativity, by contrast, in which space (more precisely, space-time) no longer has constant curvature but rather a curvature depending essentially on the distribution of matter, geometry belongs rather to the physical (and therefore synthetic) rules. Carnap's result here thus agrees completely with our argument – and also with the conception defended by Reichenbach in 1920. In the context of classical mathematical physics, Poincaré is perfectly correct: physical geometry belongs to the a priori part of our theoretical framework and hence to the conventional part. In the context of the general theory of relativity, however, Poincaré is incorrect: in this context physical geometry belongs rather to the empirical part of our theoretical framework and hence to the nonconventional part.[7]

Carnap's conception in *Logical Syntax* is thus in a much better position to establish a meaningful version of conventionalism than is a purely general Duhemian holism. Unfortunately, however, Carnap's conception has its own fatal difficulties – difficulties that have only become clear in the course of the Quinean criticism of the concept of analyticity. But this matter I must leave for Part Three.

7 For the relation between Carnap's program in *Logical Syntax* and Reichenbach's early view of the constitutive a priori, see Chapter 3, especially footnote 6 (this volume).

PART TWO

DER LOGISCHE AUFBAU DER WELT

5

CARNAP'S *AUFBAU* RECONSIDERED

Rudolf Carnap's *Der logische Aufbau der Welt,*[1] written largely in the years 1922–5 and published in 1928, is generally – and rightly – regarded as one of the most important classics of twentieth-century positivist thought. But what exactly is the importance of this great work? Precisely where does its significance lie?

The most widely accepted view of this question, I think, runs as follows. Central to twentieth-century positivism is the doctrine of verificationism – the doctrine that the cognitive meaning of all scientific statements must ultimately consist in their consequences for actual and possible sense experiences. And it is this radically empiricist doctrine, above all, that forms the basis for the notorious antimetaphysical attitude of twentieth-century positivism: in virtue of their unverifiability, metaphysical statements are deprived of all cognitive meaning as well. Yet this radically empiricist program also requires a positive construction, for one must show how the nonmetaphysical statements of science and everyday life are actually translatable into terms referring only to sense experiences. In other words, twentieth-century positivism requires a phenomenalistic reduction.

The *Aufbau,* on this reading, is important primarily for its attempt at just such a phenomenalistic reduction:

> Radical reductionism, conceived now with statements as units, set itself the task of specifying a sense-datum language and showing how to translate the rest of significant discourse, statement by statement, into it. Carnap embarked on this project in the *Aufbau.* (Quine 1951/1963, p. 39)

1 Carnap (1928a/1967); references are given in the text by section numbers.

On this reading, then, the *Aufbau* is best seen as an exceptionally detailed and rigorous attempt to execute concretely the program of Russell's (1914) *Our Knowledge of the External World*:

> To account for the external world as a logical construct of sense data – such, in Russell's terms, was the program. It was Carnap, in his *Der logische Aufbau der Welt* of 1928, who came nearest to executing it. (Quine 1969, p. 74)

Such, as I have said, is the most widely accepted view of the *Aufbau*'s significance.

It is also widely accepted, however, that the *Aufbau* fails in this reductionist project. At a crucial point – precisely where one moves from private sense experience to physical objects, in fact – the construction breaks down decisively; in particular, we are no longer presented with explicit definitions or translations at all.[2] Moreover, this failure is clearly acknowledged, with characteristic honesty and rigor, by Carnap himself.[3] From this point of view, then, the ultimate significance of the *Aufbau* – its significance for us – lies in its precise and rigorous exhibition of the failure of phenomenalistic reductionism. The *Aufbau* shows us exactly what is wrong with radical empiricism and verificationism and, therefore, prepares the way for more liberal and holistic conceptions.[4]

I think that this widely shared conception of the primary aim and significance of the *Aufbau* is fundamentally misguided. It is true, of course, that the *Aufbau* contains an important attempt at a phenomenalistic reduction. It is also true that this attempt fails. Yet focusing attention exclusively on the issue of phenomenalism leads to a serious distortion of the true philosophical context and real philosophical motivations of Carnap's work. As a result, we have distorted the philosophical context and motivations lying behind the development of twentieth-century positivism as well.

I

The *Aufbau*, as we have seen, is supposed to be first and foremost a contribution to radical empiricism. By applying the powerful new tools of modern

2 See Quine (1951/1963, pp. 39–40; 1969, pp. 76–7).

3 See the Preface to the second edition of Carnap (1928a/1967, p. vi) and Carnap (1963a, p. 19).

4 This, of course, is how Quine (1951/1963, pp. 40–2; 1969, pp. 77–84) uses the *Aufbau*. See also Putnam (1975, pp. 19–20): the significance of Carnap's work is precisely the resulting "proof," as it were, that phenomenalism is false. Goodman (1963), on the other hand, argues against the current antiphenomenalist consensus; he does appear to agree, however, that it is in connection with the issue of phenomenalism that the *Aufbau* finds its primary significance.

logic, the theory of relations and set theory, its principal aim is to give new rigor and force to traditional empiricist doctrine. But there are several obvious features of the text that do not cohere at all well with this picture.

First, much of the actual logical construction in the *Aufbau* takes place within the domain of private sense experience: the domain that Carnap calls the "autopsychological." Carnap (1928a/1967) begins with unanalyzed momentary cross sections of experience – "elementary experiences" – that are related to one another by a two-place relation *Rs* of "recollection of similarity" (§78). Using a complicated procedure of "quasi-analysis" (§§67–74), he attempts to divide the elementary experiences first into "quality classes," whose elements all agree in containing a particular sensation such as a blue spot in a given region of the visual field (§81), and then into "sense classes," which correspond intuitively to different types or modalities of sensations such as visual, auditory, tactile, and so on (§85). The next problem is to distinguish the different sense classes from one another, and this is done on the basis of dimensionality considerations: the visual field, for example, is the one and only sense modality having exactly five dimensions (§86). Carnap then proceeds to distinguish the three-dimensional color subspace (hue-saturation-brightness) of the visual field from the two-dimensional subspace of visual field places (§§88–9) and is now – and only now – in a position to talk about actual color sensations (§§90–3). It is at this point, finally, that Carnap attempts to step beyond the domain of the autopsychological into the external or physical world, in essence, by projecting color sensations onto the objects in three-dimensional space to which they correspond (§§94, 125–8).

Note that Carnap does not begin, as in much traditional empiricism, with sensations or sense data such as color patches and the like as basic or primitive elements. Under the influence of holistic and Gestalt ideas (§§67, 75), he explicitly rejects such primitive sensory "atoms," and instead arrives at concrete sensations only at the end of an intricate construction (§93). It is here, in fact, that Carnap introduces his main technical innovation: the procedure of quasi analysis, which attempts to do for similarity relations what Frege and Russell have done for equivalence relations (§§40, 70–3).[5]

5 Here also is where technical problems are likely to arise. Goodman (1951, §§V.3, V.5) raises difficulties for quasi analysis based on the possibilities of "companionship" and "imperfect community," and he criticizes Carnap for relying on questionable "extrasystematic assumptions" ruling out such possibilities. I think that these difficulties may not be as serious as Goodman takes them to be; Carnap (1928a/1967, §122) is quite explicit that his constructions are not fashioned a priori, as it were, but depend on empirical assumptions that may issue in substantial revisions of the system if false. On the other hand, Goodman (1951, §V. 6) alludes to another technical problem that does, I think, vitiate the construction: the

Yet, if Carnap's main goal is really the vindication of phenomenalistic reductionism, why should he spend so much time and technical ingenuity on an elaborate construction that takes place entirely within the domain of private experience? Why does he not simply take concrete sensations as primitive[6] and devote himself instead to a more detailed treatment of the construction of the physical world out of such sensations?

A second and more fundamental factor militating against a straightforwardly phenomenalistic reading of the *Aufbau* is this. There is no doubt that, despite the peculiarities of Carnap's procedure just noted, the *Aufbau* does present a phenomenalistic system. The construction begins from a solipsistic or autopsychological basis (§§64–6) in the private sense experience of a single individual, and it proceeds by attempting explicitly to construct everything else – first the physical world (§§125–8) and then even other individuals with their own private sense experiences (§§145–8) – from this initial autopsychological basis alone. Yet Carnap stresses repeatedly that the specific system he presents is only one possible "constructional system [*Konstitutionssystem*]" among many others (§§57–63, 122). In particular, a constructional system built on a physical rather than an autopsychological basis is equally possible and legitimate (§§57, 59, 62). It is possible, according to Carnap, to construct a phenomenalistic system in which everything is reducible to private experience; it is equally possible to construct a materialistic system in which everything – including private experience – is reducible to the objects of physics (§§57, 62).

A constructional system built on a physical basis has, in fact, important advantages over a phenomenalistic one. Such a system starts from "the only domain (namely, the physical) which is characterized by a clear regularity of process," and hence "from the standpoint of empirical science the constructional system with physical basis constitutes a more appropriate arrangement of concepts than any other" (§59). It is also true, however, that a phenomenalistic system has important advantages over a physicalistic system; the former reflects what Carnap calls "epistemic primacy," that is,

definition of *quality class* (1928a/1967, §112) presupposes that the number of elementary experiences is finite (see also §180, which states this explicitly), whereas the topological notion of *dimension number* employed in §115 presupposes that the number of elementary experiences is infinite (otherwise all "spaces" in question have zero dimension).

6 The Preface to the second edition (1928a/1967, p. vii) states that Carnap would now prefer to start with sensations or "concrete sense data" after all. This can in no way be taken as a clearer recognition and endorsement of the aims of phenomenalism, however, for, in the very next paragraph, he states even more emphatically that he would now prefer a *physicalistic* system. It is more likely that purely technical difficulties of the kind mentioned earlier in footnote 5, this chapter, are motivating Carnap here.

the order in which objects come to be known in the process of cognition (§54). Since, according to Carnap, physical objects are known through the mediation of autopsychological objects but not vice versa, it follows that an autopsychological basis is most appropriate from an epistemological viewpoint (§§59–60, 64).

Now, Carnap chooses to investigate a system with an autopsychological basis because of his "intention [*Absicht*] to have the constructional system reflect not only the logical-constructional order of the objects, but also their epistemic order" (§64). It is clear, however, that this choice [*Wahl*] is just that – and is not in any way a philosophical necessity, as it were, stemming from an antecedent commitment to phenomenalism as a philosophical doctrine.[7] Indeed, Carnap explicitly and repeatedly disclaims any such commitment (§§60, 175–8). The general discipline that Carnap is here instituting and exemplifying – the discipline of "construction theory [*Konstitutionstheorie*]" – is entirely neutral with respect to all such "metaphysical" questions (§§177–8), and one important aspect of this neutrality is that construction theory is interested equally in all possible forms of constructional system, not just in a phenomenalistic form. Thus, whereas there is no doubt that Carnap does wish to demonstrate the possibility of a phenomenalistic system in the *Aufbau*, construction theory itself has a much more general aim: to show "the possibility, in general, of a constructional system" and "the possibility, in principle, of translating all scientific statements into statements within a constructional system" (§122).

This last point is so important to Carnap that he reemphasizes it at length in his description of the writing of the *Aufbau* in his "Intellectual Autobiography" (1963a, pp. 16–20). He stresses that, for him, phenomenalism,

7 In particular, Carnap shows no interest whatever in the philosophical skepticism motivating Russell (1914, Chapter III), for example. On the contrary, Carnap's concern with "epistemic primacy" is based on nothing more than the desire to "rationally reconstruct" the actual (empirical) process of cognition (1928a/1967, §100). Thus, an important part of his motivation for starting with elementary experiences rather than "atomistic" sensations is based on the purely empirical findings of Gestalt psychology (§67); he is entirely ready to revise his constructions if the "results of the empirical sciences" make this necessary (§122); and so on. On the other hand, see Carnap (1963a, p. 50): "Under the influence of some philosophers, especially Mach and Russell, I regarded in the Logischer Aufbau a phenomenalistic language as the best for a philosophical analysis of knowledge. I believed that the task of philosophy consists in reducing all knowledge to a basis in certainty. Since the most certain knowledge is that of the immediately given, whereas knowledge of material things is derivative and less certain, it seemed that the philosopher must employ a language which uses sense-data as a basis." It is remarkable that there is no trace at all of such concern for philosophical certainty in the *Aufbau* itself, however. For further discussion of such retrospective passages from Carnap's "Intellectual Autobiography," see Chapter 6, §IV (this volume).

materialism, and so on are merely so many "ways of speaking," representing nothing more than pragmatically motivated choices of language, and he asserts the neutral character of construction theory most explicitly:

> When I developed the system of the *Aufbau*, it actually did not matter to me which of the various forms of philosophical language I used, because to me they were merely modes of speech, and not formulations of positions. . . . The system of concepts was constructed on a phenomenalistic basis . . . . However, I indicated also the possibility of constructing a total system of concepts on a physicalistic basis. The main motivation for my choice of a phenomenalistic basis was the intention to represent not only the logical relations among the concepts but also the equally important epistemological relations. The system was intended to give, though not a description, still a rational reconstruction of the actual process of the formation of concepts . . . . The ontological theses of the traditional doctrines of either phenomenalism or materialism remained for me entirely out of consideration. (p. 18)

It is therefore clear beyond the shadow of a doubt, I think, that the *Aufbau* has a much more general aim than the particular construction of a phenomenalistic system.

Yet there is a strong temptation to distrust such Carnapian claims to ontological neutrality, a temptation that stems, I think, from the idea that the antimetaphysical attitude of the positivists must rest ultimately on verificationism and radical empiricism. From this point of view, a phenomenalistic system must have a central and privileged place after all, and Carnap's attempt to distance himself from traditional phenomenalism must be seen as a sham. Perhaps, however, it is not Carnap, but rather the idea that an antimetaphysical attitude must rest on radical empiricism, that is at fault here. Indeed, in view of the fact that Carnap persists in his antimetaphysical attitude long after he explicitly acknowledges the failure of phenomenalistic reductionism and radical empiricism, this latter alternative should appear much more plausible. My own view is that Carnap's antimetaphysical attitude is not, in the end, based on empiricist doctrines at all, but rather on precisely the attempt to find a peculiarly philosophical vantage point that is neutral with respect to all traditional metaphysical disputes. That is, Carnap does not ultimately reject the metaphysical tradition on crudely verificationist grounds, but rather because he thinks he has found a replacement – a "scientific" replacement – for metaphysics.[8]

8 Compare (1963a, pp. 18–19): "This neutral attitude toward the various forms of language, based on the principle that everyone is free to use the language most suited to his purposes,

II

If the primary aim of the *Aufbau* is not the construction of a particular form of phenomenalistic system – if the aim of construction theory is really much more general – what then is this aim?

In §1, Carnap (1928a/1967) explains that the aim of construction theory is to attempt

> a step-by-step derivation or "construction [*Konstitution*]" of all concepts from certain fundamental concepts, so that a genealogy of concepts results in which each one has its definite place. It is the main thesis of construction theory that all concepts can in this way be derived from a few fundamental concepts, and it is in this respect that it differs from most other ontologies [*Gegendstandstheorien*].

This method, as he explains in §2, will lead to the goal of a unified science. But why is *this* so important? At the end of §2, we find the following cryptic remark:

> Even though the subjective origin of all knowledge lies in the contents of experience and their connections, it is still possible, as the constructional system will show, to advance to an intersubjective, *objective world*, which can be conceptually comprehended and which is the same for all observers.

This remark, in my view, encapsulates the most fundamental aim of the *Aufbau*, namely, the articulation and defense of a radically new conception of objectivity.

Carnap's conception of objectivity emerges in the next several sections, where it is explicitly connected with the notion of logical form or structure. Thus, in §6, Carnap states his goal this way: "It will be demonstrated that it is in principle possible to characterize all objects through merely structural properties (i.e., certain formal-logical properties of relation extensions or complexes of relation extensions) and thus to transform all scientific statements into purely structural statements." Section 10 begins: "In the following, we shall maintain and seek to establish the thesis that science deals only with the description of structural properties of objects." The connection between logical form or structure and the notion of objectivity is made most

has remained the same throughout my life. It was formulated as 'principle of tolerance' in *Logical Syntax* and I still hold it today. . . . [I]f one proceeds from the discussion of language forms to that of the corresponding metaphysical theses about the reality or unreality of some kind of entities, he steps beyond the bounds of science."

explicitly in §16:

> each scientific statement can in principle be transformed into a statement which contains only structural properties and the indication of one or more object domains. Now, the fundamental thesis of construction theory (cf. §4), which we will attempt to demonstrate in the following investigation, asserts that fundamentally there is only one object domain and that each scientific statement is about the objects in this domain. Thus, it becomes unnecessary to indicate for each statement the object domain, and the result is that *each scientific statement can in principle be so transformed that it is nothing but a structure statement.* But this transformation is not only possible, it is imperative. For science wants to speak about what is objective, and whatever does not belong to structure but to the material (i.e., anything that can be pointed out in a concrete ostensive definition) is, in the final analysis, subjective.... From the point of view of construction theory, this state of affairs is to be described in the following way. The series of experiences is different for each subject. If we want to achieve, in spite of this, agreement in the names for the entities which are constructed on the basis of these experiences, then this cannot be done by reference to the completely divergent content, but only through the formal description of the structure of these entities.[9]

Thus, for Carnap, construction theory, the unity of science (what he calls "the unity of the object domain" in §54), logical form or structure, and scientific objectivity are intimately connected.[10] Our problem is to understand the nature and significance of this connection.

9 See also §66: "Since the stream of experience is different for each person, how can there be even one statement of science which is objective in this sense (i.e., which holds for every individual, even though he starts from his own individual stream of experience)? The solution to this problem lies in the fact that, even though the *material* of the individual streams of experience is completely different, or rather altogether incomparable, since a comparison of two sensations or two feelings of different subjects, so far as their immediately given qualities are concerned, is absurd, certain *structural properties* are analogous for all streams of experience. Now, if science is to be objective, then it must restrict itself to statements about such structural properties, and, as we have seen earlier, it can restrict itself to statements about structures, since all objects of knowledge are not content, but form, and since they can be represented as structural entities (see §15 f.)."

10 Carnap, like Schlick (see Chapter 1, footnote 20, this volume), sees the abstract formulations of contemporary mathematical physics as paradigmatic of the kind of "structural objectivity" he is attempting to capture. Thus, in the ellipsis of the preceding quotation from §16, Carnap has: "We easily notice this de-subjectification in *physics*, which has already transformed almost all concepts of physics into pure stuctural concepts. In the first place, all mathematical concepts are reducible to concepts of the theory of relations; four-dimensional tensor or vector fields are stuctural schemata; the network of world-lines with the relations of coincidence and proper time is a structural schema in which only one or two relations are still designated by names, which, however, are also already univocally determined by the character of the schema."

Carnap introduces the concept of the form or structure of a relation in §11. The structure of a relation is the class of all relations that are isomorphic to it, or, what comes to the same thing, the totality of its formal properties, such as symmetry, reflexivity, transitivity, connectedness, and so on, that can be expressed using only purely logical notions. The underlying idea, of course, is that two different relations – relations that differ in "content [*inhaltlicher Sinn*]" – such as the relations of "later-than" defined between moments of time and "being-to-the-right-of" defined between points on a line, for example, may have exactly the same logical form or structure (in this case, both relations are continuous linear orderings). So far, then, the idea is a perfectly standard and familiar part of the modern theory of relations.

In the next several sections, however, Carnap, argues that only the logical form or structure of a relation is objectively or scientifically communicable: any excess "content" going beyond logical structure must rest ultimately on ostensive definitions, and these, according to Carnap, provide no intersubjective meaning. For truly objective communication, then, we must require that all relations are given only through descriptions of their structure – through what Carnap calls "purely structural definite descriptions" (§§12–15). Thus, "within any object domain, a unique system of definite descriptions is in principle possible, even without the aid of ostensive definitions.... any intersubjective, rational science presupposes this possibility" (§13); and "*a definite description through pure structure statements is generally possible to the extent to which scientific discrimination is possible at all*" (§15).

This last idea appears highly paradoxical, of course, for the notion of logical form or structure is such that two different relations, such as the temporal order and the spatial order on a line, may have precisely the same formal structure. How, then, can two such relations be discriminated from one another solely on the basis of structure? Here is where the "fundamental thesis of construction theory" comes into play: if we imbed all relations within a single global structure of relations, then we may hope to be able to discriminate (according to Carnap, we *must* be able to discriminate) formally identical relations through their differing formal "places" within this all-encompassing global structure (see §§14–15). Thus, for example, whereas the temporal order and the spatial order on a line are "locally" structurally identical (both are continuous linear orderings), the latter occurs as a subspace of the total three-dimensional spatial order whereas the former does not; moreover, within the global space-time manifold, the temporal dimension is itself formally distinguishable from the three spatial dimensions; and so on.

This is why the unity of science – the unity of the object domain (§4) – is so important to Carnap. It is only if all concepts are part of a single interconnected system of concepts that we can hope to do what, according to Carnap's new conception of scientific objectivity, we must do: discriminate all concepts from one another solely on the basis of their purely formal or structural properties. This also explains the, at first sight, extremely peculiar method of definition that Carnap actually employs in the *Aufbau*. Beginning with the two-place relation *Rs* of recollection of similarity, he divides the elementary experiences first into quality classes and then into sense classes or sense modalities. Each sense modality is ordered by a two-place similarity relation, which, according to Karl Menger's topological definition of dimensionality, can be assigned a dimension number.[11] The visual field then is picked out from all other sense modalities by the purely formal properties of its associated dimension number: namely, the visual field, according to Carnap, is the unique sense modality having exactly five dimensions (§86; all other sense modalities have either two, three, or four dimensions). We then pick out the three-dimensional color subspace of the visual field from the two-dimensional subspace of visual field places by similar purely formal considerations (§§88–91); but, even so, we are not yet in a position formally to define the various individual colors. This, in fact, cannot be done until after we have constructed the physical world, wherein individual colors can be defined as the colors of various types of physical things: green is the color of foliage, for example (§134). Thus, Carnap does not begin his construction with sensations or sense data precisely because even these entities must ultimately be defined on the basis of purely formal or structural properties – by their logical "places" within a single interconnected system of concepts.

Viewed from this perspective, Carnap's project has less affinity with traditional empiricism and more with Kantian and neo-Kantian conceptions of knowledge.[12] The primary problem is to account for the objectivity of scientific knowledge, and the method of solution is based on a form/content distinction. Scientific knowledge is objective solely in virtue of its formal or structural properties, and these properties are expressed through the "places" of items of knowledge within a single unified system of knowledge. The project is not *strictly* Kantian, of course, because the notion of form or structure in question here is a purely logical one, understood solely in

11 See Carnap (1954/1958, §46c). Compare footnote 5, this chapter, however.

12 For the earliest recognition of such Kantian and neo-Kantian influences on the *Aufbau* see Haack (1977), Moulines (1985), and Sauer (1985).

terms of formal logic. For Kant himself, merely formal logic is quite inadequate for the constitution of objectivity, and we need to supplement it with a "transcendental logic" that makes essential reference to intuition: the "pure intuitions" of space and time. Now, in the context of the much more powerful conception of formal logic bequeathed to him by Frege and Russell, Carnap finds such an independent appeal to the "forms of intuition" quite unnecessary,[13] and space and time have no special status: they simply find their proper places in the constructional system along with all other concepts (§§87, 124–5). In other words, whereas Carnap retains the Kantian connections among objectivity, the notion of form or structure, and the a priori (for formal logic is itself certainly a priori for Carnap), he now has no need whatever for Kant's *synthetic* a priori.

Nevertheless, it is of the utmost importance that Carnap's conception of knowledge and meaning is Kantian – and in fact quite opposed to traditional empiricism – in that it is "holistic" rather than "atomistic."[14] Concepts do not derive their meaning "from below" – from ostensive contact with the given. Indeed, such merely ostensive contact with the given is the very antithesis of truly objective meaning and knowledge; for objective meaning can only be derived "from above" – from formal or structural relations within the entire system of knowledge. Such a formalistic and holistic conception of meaning and knowledge was in fact widely held throughout the period – and held by thinkers who are generally regarded as paradigmatically empiricist. The conception plays a prominent role, for example, in Moritz Schlick's (1918) *General Theory of Knowledge* and other early writings,[15] in Russell's (1927) *The Analysis of Matter*,[16] and in C. I. Lewis's (1929) *Mind and the World-Order*.[17] All of these works disengage meaning and knowledge from ostension, and

13 Note that Carnap's project for characterizing concepts purely formally or structurally only begins to make sense in the context of polyadic logic or the modern theory of relations: in traditional monadic or syllogistic logic, concepts have only a single formal property, namely, cardinality. For an attempt to articulate the significance of the distinction between monadic and polyadic logic for Kant's conception of geometry and pure intuition, see again Friedman (1992a, Chapters 1 and 2).

14 For a somewhat different reading of Carnap's holistic approach to meaning in the *Aufbau*, see Coffa (1991, Chapter 11, pp. 218–22).

15 Schlick (1918/1984, 1926/1979). See Chapter 1 (this volume) for a discussion of Schlick on this point.

16 See Demopoulos and Friedman (1985/1989).

17 See especially Chapters III–V of Lewis (1929), e.g., pp. 120–1: "That there is direct apprehension of the immediate, it would be absurd to deny; but confusion is likely to arise if we call it 'knowledge.' There are no 'simple qualities' which are named by any name; there is no concept the denotation of which does not extend beyond the immediately given, and beyond what *could be* immediately given. And without concepts, there is no knowledge."

lodge them instead in the system of logical relationships among our concepts. All of these writers, including Carnap, are clearly indebted to the notion of "implicit definition" deriving from Hilbert's (1899) axiomatization of Euclidean geometry.[18] What distinguishes Carnap from the others is simply his rigorously constructive spirit: he transforms holistic and formalistic sentiments into a definite technical program for characterizing all concepts of science through purely structural definite descriptions formulated within a single constructional system.

This perspective on the *Aufbau* illuminates the two features of the text noted in §I of this chapter. It is clear, first of all, that logical construction is just as important within the purely private domain of the autopsychological as it is anywhere else. For this domain, too, has a formal or logical structure, and construction theory has the task of revealing all such structure. Indeed, in view of the fact that the domain of the autopsychological is, at first sight, the domain where purely ostensive meaning has a natural and proper place, the demonstration that even this domain can be characterized through its logical structure alone is especially important to Carnap; for only so do we see how scientific objectivity extends to *all* of our concepts.[19] From this point of view, then, the step leading from the private domain of the autopsychological to the external domain of the physical has no special importance, and this explains why Carnap devotes so much more space and ingenuity to the former domain.

By the same token, it is also clear that other constructional systems besides the particular one Carnap attempts to construct here are equally important and legitimate. For any such system contributes equally well to the goal of revealing the logical structure of, and logical relations among, our concepts. What construction theory primarily seeks is a characterization of all concepts through their formal or structural properties, and, as we have seen, what this requires is the unity of science – the unity of the object domain.

18 Schlick (1918/1985, §7) endorses Hilbertean implicit definitions; Lewis (1927, Chapter III) expounds a "relational" conception of meaning that is clearly indebted to Hilbert; Russell (1927, pp. 136–7) defends implicit definitions using an example derived from Eddington; Carnap endorses Hilbertian implicit definitions (and refers to Schlick) in §15. (Note added in 1998: As Richardson [1998, pp. 40–7] rightly points out, it is inaccurate to characterize Carnap as "endorsing" implicit definitions here. One of the main points of §15, in fact, is to distinguish his own use of *explicit* definitions – i.e., purely structural definite descriptions – from the merely *implicit* definitions favored by Schlick. In particular, Carnap turns implicit definitions into explicit ones by adding a *uniqueness* clause.)

19 This extension of logical structure to the domain of the autopsychological itself marks an important point of contrast between Carnap's conception of objectivity and Schlick's in *General Theory of Knowledge*. For discussion, see the Postscript to Chapter 1 (this volume).

The choice of one particular object domain over others – a phenomenalistic domain in the autopsychological realm of private experience over a materialistic domain in the primitive entities of physics, for example – is then a matter of comparatively little significance. Scientific objectivity, according to Carnap, requires precisely such a unified system of purely structural definite descriptions; a phenomenalistic reduction of all concepts to the given is in no way essential.

III

If we are correct in our interpretation of the primary aim of the *Aufbau*, that is, the general goal of construction theory, then it follows that the failure of phenomenalistic reductionism cannot be the most fundamental problem facing the *Aufbau*. The real problems are correspondingly more general and, I think, deeper.

The aim of construction theory is the characterization of all concepts of science through purely structural definite descriptions. We have briefly sketched the method by which the *Aufbau* attempts to achieve this aim above. Starting from the two-place relation *Rs* defined between elementary experiences, Carnap constructs the general class of sense modalities by quasi analysis, and then picks out the visual field from all other sense modalities by means of its unique dimensionality. This construction then becomes the fixed point from which all other concepts are generated: first color classes and color sensations, then physical objects, and finally even other experiencing subjects with their own elementary experiences. What makes the entire system of purely structural definite descriptions possible, in other words, is the fact (according to Carnap) that the visual field is the unique sense class based on *Rs* having exactly five dimensions (see §§115, 119).

Yet the basic relation *Rs* is so far itself undefined: it is simply introduced as a nonlogical primitive (see §§108, 119). We have, to be sure, drastically reduced the number of nonlogical primitives and have characterized almost all concepts purely formally or structurally. But the ultimate goal of construction theory still eludes us, for the scientific objectivity of the basic relation *Rs* has itself not yet been shown. Moreover, since all other concepts have been reduced to *Rs*, all we have really shown so far is that they are objective if it is. In other words, Carnap's program requires a *complete* formalization of all concepts of science, and we have achieved so far merely a partial (albeit still very impressive) formalization. Carnap raises this problem, and attempts to solve it, in §§153–5 (which are innocently labeled "may be omitted"). Thus:

> A purely structural statement must contain only logical symbols; in it must occur no undefined basic concepts from any empirical domain. Thus, after the constructional system has carried the formalization of scientific statements to the point where they are merely statements about a few (perhaps only one) basic relations, the problem arises whether it is possible to complete the formalization by *eliminating from the statements of science these basic relations* as the last non-logical objects. (§153)

As Carnap indicates, this problem will arise for any constructional system regardless of the choice of nonlogical primitive(s), and it has no intrinsic connection with the choice of a phenomenalistic system.

How is it possible to eliminate even the primitive nonlogical concepts from a constructional system? The method that suggests itself to Carnap is again the method of purely structural definite description. In constructing other objects from our nonlogical primitive(s), we will make essential use of certain empirical facts. In Carnap's system, for example, we make essential use of the (putative) fact that there is one and only one sense modality based on *Rs* that is exactly five-dimensional. Now, Carnap claims, by ascending to high enough levels in our constructional system, we may hope to find sufficient such empirical facts to uniquely characterize our chosen basic concept(s) in contradistinction to all other possible basic concepts (§153). We could define *Rs*, for example, as the unique basic relation such that there is one and only one sense modality based on it having exactly five dimensions.[20]

What we have done, in effect, is eliminate the constant relation *Rs* in favor of a variable ranging over relations: *Rs* is the unique relation satisfying certain empirical conditions. But a final difficulty now arises. For, unless we in some way restrict the domain of relations over which our variable is ranging, the uniqueness claim in question will be generally false. Given one basic relation (or set of basic relations) satisfying the empirical conditions,

> [a]ll we have to do is carry out a one-to-one transformation of the set of basic elements into itself and determine as the new basic relations those relation extensions whose inventory is the transformed inventory of the original basic relations. In this case, the new relation extensions have the same structure as the original ones (they are "isomorphic"). (§154)

The difficulty also can be put in another way: assuming that our chosen empirical conditions are themselves logically consistent, the existence claim implicit in our definition of the basic relation(s) will be a logico-mathematical

20 Compare §155: Carnap himself uses a closely related fact concerning the three-dimensionality of the color solid, which is introduced as an "empirical theorem" in §§118–19.

truth, and the uniqueness claim will, in general, be a logico-mathematical falsehood.[21]

Carnap responds, then, precisely by restricting the range of our variable: we are not to consider all relations – which, as mere mathematical sets of pairs, may be "arbitrary, unconnected pair lists" – but we are to restrict ourselves to "experienceable [*erlebbaren*], 'natural' relations" or what Carnap calls "*founded*" relations (§154). Carnap next makes the extraordinary suggestion that this notion of *foundedness* may itself be considered a basic concept of logic (§154), and he completes the "elimination of the basic relation" thusly: *Rs* is the unique *founded* relation satisfying the chosen empirical conditions (§155)! The fundamental aim of construction theory has now – and only now – been reached: all concepts of science have been characterized through purely structural definite descriptions.

It is clear that, in the context of the basic motivations of the *Aufbau*, Carnap's suggested solution to the difficulty can in no way be satisfactory. We are motivated to pursue a program of complete formalization by a conception of scientific objectivity that seeks to disengage objective meaning entirely from ostension. We now find that to reach our goal we need to introduce the class of *founded* relations as a primitive notion of logic, where the *founded* relations are just the "experienceable, 'natural' relations." But what can the "experienceable, 'natural' relations" be except precisely those relations somehow available for ostension? Our original motivations, in other words, have been totally undermined by Carnap's final move. It is also clear, however, that the difficulty is an extremely fundamental one. If we succeed in disengaging objective meaning and knowledge from ostension and lodge them instead in logical form or structure, then we run the risk of divorcing objective meaning and knowledge from any relation to experience or the empirical world at all. We run the risk, that is, of erasing completely the distinction between empirical knowledge and logico-mathematical knowledge. (In these terms, Carnap's suggestion for

21 As Anil Gupta and Mark Wilson have emphasized to me, this assertion depends on assumptions about the structure of the underlying type-theoretic system with which Carnap is operating. I assume, in particular, that there are an infinite number of individuals of lowest type (axiom of infinity) and that we are working either in the simple theory of types or in the ramified theory *with* the axiom of reducibility. Although Carnap is very casual about the exact structure of his underlying type-theoretic system here, he does claim to capture all of classical mathematics (§107) including *n*-dimensional real-number space (§125); the above two assumptions are therefore entirely reasonable. (In addition, Gupta has reminded me that there are relations – such as the identity relation – that are explicitly definable in our underlying type-theoretic system. For such relations, both the existence claim and the uniqueness claim are logical truths.)

introducing the notion of *foundedness* may be seen as an attempt to evade the problem simply by counting *empirical* or *nonlogical* as itself a basic concept of logic.)

The importance and significance of this problem can perhaps best be brought out by seeing how it arises for other thinkers of the period.[22] Moritz Schlick provides a particularly clear example. In *General Theory of Knowledge*, as briefly noted earlier, we also are presented with a formalistic or holistic conception of objective meaning. Precise and rigorous meanings cannot be based on images or sense data, for these are both fleeting and irreducibly particular – no truly general representation can arise from a sensory presentation. Schlick concludes that objective meaning can have no dependence whatsoever on intuition, and it must be given instead by implicit definitions that completely characterize a concept solely in virtue of its logical relationships to all other concepts. Yet Schlick also immediately observes that a serious problem then arises, for implicit definitions establish no connection between thought and experience at all:

> in implicit definition we have found a tool that makes possible completely determinate concepts and therefore rigorously exact thought. However, we require for this purpose a radical separation between concepts and intuition, thought and reality. To be sure, we place the two spheres one upon the other, but they appear to be absolutely unconnected, the bridges between them are demolished.[23]

Indeed, as Schlick also observes, implicit definitions require only the logico-mathematical consistency of the system of judgments in question, so that empirical truth has so far been subject to no further constraints beyond those of logico-mathematical truth. We are faced, in other words, with the clear possibility of a collapse into idealism and the coherence theory of truth.

Schlick struggles with this problem throughout his philosophical career, but never achieves a coherent solution. In the late 1920s and early 1930s,

22 A problem exactly parallel to Carnap's problem of §§153–5 arises for Russell's (1927) notion of "purely structural knowledge," and it is explicitly raised, in fact, by the mathematician M. Newman: see Demopoulos and Friedman (1985/1989) for details.

23 Schlick (1918/1985, §7). Note that what has been "demolished" here is precisely Kant's pure intuition. For Kant, pure spatiotemporal intuition is simultaneously a vehicle for precise mathematical reasoning and a "form" within which we experience nature through the senses. For Kant, then, there is a *necessary* connection between rigorous mathematical reasoning and experience. Hilbert's axiomatization of geometry – along with other nineteenth-century foundational developments – then frees mathematical reasoning from any connection whatever with spatiotemporal intuition; unfortunately, however, the necessary relation to our experience of nature has been entirely dissolved as well. This, in the end, is the source of our problems here.

during the heyday of the Vienna Circle, he articulates a classically empiricist (or "atomistic") position according to which all empirical meaning is based on ostensive definitions after all; and this, in fact, is how he arrives at the verifiability theory of meaning. This move certainly succeeds in distinguishing empirical propositions from logico-mathematical propositions, yet it is also totally at variance with our earlier insight that all objective meaning must rest in the end on logical form or structure. Schlick (1932b) acknowledges the resulting conflict in "Form and Content," but is again unable to find a coherent resolution.[24] His struggles with the form/content distinction are nonetheless of the utmost importance in clearly exhibiting the intellectual temptations that actually give rise to the verifiability theory of meaning, temptations that are so strong that even Carnap momentarily succumbs to them in 1928.[25] Fortunately, however, Carnap's rigorous method of thought immediately suggests a much more fundamental approach to the problem.

To understand fully the path Carnap ultimately takes, it is useful to take a preliminary brief look at Wittgenstein's (1922) *Tractatus*. The *Tractatus* presents a view that has close affinities with the holistic and formalistic conception of meaning that we have been considering: the sense of a proposition is identified with its "place" in "logical space" (3.4–3.42). Moreover, we also find an idea that appears to be very close indeed to Carnap's strategy of finding purely structural definite descriptions for all concepts:

> We can describe the world completely by means of fully generalized propositions, that is, without first correlating any name with a particular object.
>
> Then, in order to arrive at the customary mode of expression, we simply need to add, after an expression like 'There is one and only one x such that . . . ,' the words: 'and that x is a.' (5.526)

Finally, the discussion of causality and mechanics at 6.3–6.3751 even suggests a justification for this kind of strategy that is also close to Carnap's: the possibility of purely structural definite descriptions rests on sufficient *de facto* asymmetries in the phenomena; therefore, we may presuppose that

24 See Chapter 1 (this volume) for further discussion. See especially the Postscript to Chapter 1 for Schlick's attempt to rebuild the bridge between thought and experience in the period of *General Theory of Knowledge*. And see also the conclusion of this Postscript for a comparison of Schlick and Carnap on the problem of relating thought and experience, in light of which (Chapter 1, footnote 30, this volume) I no longer view the difficulty of §§153–5 of the *Aufbau* as such a fundamental one.

25 Carnap (1928b/1967). For an accurate representation of the role of *Pseudoproblems* in Carnap's thought, see Carnap (1963): in §I, "The Development of my Thinking," it is not mentioned at all; it receives only a single brief mention on p. 46 of §11, "Philosophical Problems."

the laws of nature – in so far as nature is thinkable – will display such formal asymmetries.[26]

Now it would, of course, be extremely rash confidently to ascribe an *Aufbau*-style program to the *Tractatus.* Yet it is essential, in any case, to see that Carnap's problem of §§153–5 cannot arise in the *Tractatus.* The problem was that the purely structural definite description that Carnap introduces to eliminate the basic relation threatens to turn into either a logico-mathematical truth or a logico-mathematical falsehood, depending on whether we focus on the existence claim or the uniqueness claim implicit in that definition. But this in turn depends crucially on what exactly we mean by "logico-mathematical truth." In particular, it is only if our underlying logic contains a sufficient amount of what we now call set theory – the power set axiom and the axiom of infinity, for example – that the problem arises, for the problem depends precisely on our ability to prove the existence of "too many" relations. Carnap is apparently willing to count virtually all of set theory as logical, and this is why the problem arises for him.[27]

The *Tractatus*, on the other hand, emphatically rejects set theory (6.031) and instead articulates an extremely restricted conception of logico-mathematical truth apparently limited to something in the vicinity of primitive recursive arithmetic (6.2–6.241).[28] On this conception Carnap's problem will then not arise, for neither the existence claim nor the uniqueness claim implicit in the definition of the basic relation will be a logico-mathematical truth (falsehood). If true at all, such claims can only be empirical – that is, nonlogical – truths. Hence, if we were to adopt a Tractarian conception of logic and mathematics, Carnap's strategy of complete formalization could perhaps be successfully carried out after all. But we would, of course, have paid a terrible price for this success: the total emasculation of classical mathematics.

26 See Wittgenstein (1922, 6.36–6.3611, especially 6.3611): "when people say that neither of two events (which exclude one another) can occur, because there is nothing to cause the one to occur rather than the other, it is really a matter of our being unable to describe *one* of the two events unless there is some sort of asymmetry to be found. And *if* such an asymmetry *is* to be found, we can regard it as the *cause* of the occurrence of the one and the non-occurrence of the other." Compare §§12–16 of the *Aufbau.*

27 See footnote 21, this chapter.

28 Speaking of logico-mathematical truth in the context of the *Tractatus* is actually rather misleading, for Wittgenstein shows no interest whatever in a logicist reduction of mathematics to logic. His conception of logic can perhaps be approximated by ramified type-theory (3.331–3.334, 4.1273) *without* the axioms of infinity (5.535) and reducibility (6.1232, 6.1233). His conception of mathematics, on the other hand, appears to be a purely combinatorial one limited to some subsystem of primitive recursive arithmetic (6.2–6.241).

The final chapter to our story is written in Carnap's next great work, *The Logical Syntax of Language* (1934c)[29] – which, more than any other of Carnap's works, is written under the explicit influence of Wittgenstein's *Tractatus*. In particular, *Logical Syntax* also gives pride of place to the "*combinatorial analysis* . . . of finite, discrete, serial structures" (§2), that is, to primitive recursive arithmetic. Yet Carnap radically transforms this conception by distinguishing, as Wittgenstein never would, between object language and meta-language.[30] Thus, logic in the sense just defined is understood as *logical syntax*, a neutral metadiscipline within which we can formulate and investigate the formal rules of any and all object languages or linguistic frameworks. The point is that, although the metadiscipline of logical syntax has itself a very restricted (and therefore uncontroversial) logical structure, we can nonetheless use it to study the logical structures of much richer (and more controversial) object languages: in particular, the language of classical mathematics and mathematical physics. We hope thereby to avoid the emasculating effects of the *Tractatus*.

Logical Syntax also follows the *Aufbau* in articulating a formalistic and holistic conception of objective meaning. Objective meaning is entirely determined by the purely formal – purely syntactical – rules of a given language or linguistic framework, and there is absolutely no question remaining concerning "content" or "interpretation" (§62). In other words, the objective meaning of an expression is a function solely of its purely formal behavior in the context of a single language, where "formal" now means "syntactical." How, then, do we avoid the problem of §§153–5 of the *Aufbau*? How do we distinguish empirical truth from logico-mathematical truth? The meanings of all expressions – both logico-mathematical expressions such as the primitive signs of logic and arithmetic and empirical expressions such as the primitive signs of physics – do indeed depend entirely on their purely formal or syntactic behavior. Yet we can still make a distinction (for any given language) on purely formal or syntactical grounds between *logical* expressions and *descriptive* (empirical) expressions (§50) and, accordingly, between *analytic* truths such as the axioms of logic and arithmetic and *synthetic* truths such as the laws of physics (§§51–2). In this way, by making the crucial move into the metadiscipline of logical syntax, Carnap has finally reached a position where he can hope to implement coherently all the elements of his underlying philosophical vision: a rigorous version of a purely formal or

29 Carnap (1934c/1937); references are given in the text by section numbers.

30 For the logical problems involved in articulating this distinction, see Goldfarb (1979); for Wittgenstein, in particular, see also Ricketts (1985).

structural conception of objective meaning and knowledge together with a rigorous, and also purely formal, distinction between logical and empirical truth. Unfortunately, even this position, too, proves to be fundamentally unstable; but that is a different story.[31]

IV

Our discussion has, I hope, raised significant problems for a straightforwardly phenomenalistic reading of the *Aufbau.* The primary aim of construction theory is not the articulation and defense of phenomenalistic reductionism; the primary problem for construction theory does not arise from the failure of phenomenalistic reductionism. What, then, is the real role of phenomenalism – and, more generally, of empiricism – in the *Aufbau*? How, in particular, does it function in the criticism of metaphysics?

First, there is no doubt at all that the *Aufbau* does defend empiricism and phenomenalism. Carnap (1928a/1967) calls the latter doctrine "subjective idealism" and clearly asserts that "[c]onstruction theory and subjective idealism agree with one another that statements about objects of cognition can, in principle, all be transformed into statements about structural properties of the given" (§177). Moreover, if the choice is between "rationalism" and "empiricism," Carnap's preference is also perfectly clear:

> Since, according to construction theory, each statement of science is at bottom a statement about relations that hold between elementary experiences, it follows that each substantive (i.e., not purely formal) insight goes back to experience. Thus, the designation "empiricism" is more justified. (§183)

Finally, Carnap even articulates a version of what will later become the verifiability principle:

> From a logical point of view, however, statements which are made about an object become statements in the strictest scientific sense only after the object has been constructed, beginning from the basic objects. For, only the construction formula of the object – as a rule of translation of statements about it into statements about the basic objects, namely, about relations between elementary experiences – gives a verifiable meaning [*verifizierbaren Sinn*] to such statements, for verification means testing on the basis of experiences. (§179)

31 See Part Three (this volume).

Indeed, since according to Carnap the number of elementary experiences is finite, each scientific statement can, in principle, be decided on the basis of experience in a finite number of steps (§180)!

Yet it is equally clear, I think, that these empiricist doctrines do not play an essential role in Carnap's criticisms of traditional metaphysics. Carnap addresses such issues in Part V of the *Aufbau*, entitled "Clarification of some philosophical problems on the basis of construction theory." In each case he distinguishes a "constructional" from a "metaphysical" version of the problem and argues that the former can be formulated within construction theory – and therefore within "rational science" – whereas the latter cannot. The discussion of the "problem of reality" (§§170–8) is developed in the most detail and, I think, is most representative of Carnap's attitude.

Section 170 introduces the "constructional" or "empirical" concept of reality – applied, in the first instance, to "physical bodies":

> These bodies are called real if they are constructed as classes of physical points which are located on connected bundles of world lines and are placed within the all-comprehending four-dimensional system of the space-time world of physics.

In other words, real physical bodies – unlike objects of dreams, hallucinations, and so on – all fit together determinately in accordance with the laws of physics. Section 171 then extends this idea to all other objects: they too are called real when they fit together determinately with others in a single law-governed system (e.g., the system of my psychological states). Thus: "*Every real object belongs to a comprehensive system which is governed by regularities*" (§171).

The "metaphysical" concept of reality – which, according to Carnap, is alone at issue in the dispute between "realism, idealism, and phenomenalism" – is characterized as "*independence from the cognizing consciousness*" (§175). Carnap argues in §176 that this concept "does not belong within (rational) science" because no notion of "independence from consciousness" suitable to the needs of the dispute "can be constructed." This emphatically does not mean, however, that the notion is metaphysical because it cannot be constructed within a phenomenalistic system; rather, according to Carnap, it cannot be constructed within *any* of the systems considered by construction theory:

> It must be noted that this [failure of constructibility] holds, not only of a constructional system which has the system form represented in our outline, but for any cognizable [*erkenntnismäßige*] constructional system, even for a system which does not proceed from an autopsychological basis, but from the experiences of all subjects or from the physical. *The* [metaphysical] *concept of*

> *reality cannot be constructed in a cognizable constructional system; this characterizes it as a nonrational, metaphysical concept.*[32]

In other words, the metaphysical concept of reality lies outside the boundaries of science, not simply because it has no experiential or verifiable meaning, but because it has no "constructional" meaning at all: that is, it has no "logical," "rational," "non-intuitive" – i.e., formal – meaning (§182).[33]

Once again, therefore, Carnap's standpoint is much more general than phenomenalism – or even empiricism. Indeed, the "metaphysical" question of reality is ultimately dissolved, for Carnap, not by ruthless application of the verifiability principle, but by the fact that construction theory itself captures the meaningful core, as it were, shared by all parties to the dispute. In particular, construction theory agrees with "realism," "idealism," "phenomenalism," and even "transcendental idealism" on all "assertions" (§177); and none of these doctrines has a privileged status. On the contrary,

> *the so-called epistemological schools of realism, idealism and phenomenalism agree within the field of epistemology. Construction theory represents the neutral foundation which they have in common. They diverge only in the field of metaphysics, that is to say (if they are meant to be epistemological schools of thought), only because of a transgression of their proper boundaries.* (§178)

It is metaphysical neutrality rather than radical empiricism that is of the essence of Carnap's position.[34]

Nevertheless, Carnap does, for a time, continue to espouse radical empiricism. Moreover, as briefly noted earlier, Carnap does, of course, adopt

32 In what appears to be a rare slip, George translates "*erkenntnismäßig*" as *experiential* here (and a similar problem occurs in §182). Since §59, for example, uses "*erlebnismäßig*" for *experiential,* and since Carnap's point here is that a constructional system need *not* have an experiential basis, something like "cognizable" or "suitable for the representation of cognitions" seems much more appropriate. There is a serious problem of what *erkenntnismäßig* actually excludes, of course; but it will not do, I think, to take it to mean *translatability into a phenomenalistic system,* for example. For Carnap uses conformity to law as the "constructional" criterion of reality (§§170–1) and holds that a *physicalistic* system is most suitable for representing conformity to law (§59).

33 Section 182 cites Schlick (1926/1979), where "rational," "formal" knowledge is explicitly contrasted with "intuitive metaphysics." (In reference to the question of translation raised in footnote 32, this chapter, one should note that Schlick also explicitly opposes *erkennen* and *erleben.*)

34 Section 178 approvingly quotes Gätschenberger: "All philosophers are correct, but they express themselves with varying degrees of ineptness, and they cannot help this, since they use the *available* language and consequently speak in a hundred sublanguages, instead of inventing one pasigraphy." Carnap concludes: "This neutral language is the goal of construction theory."

the verifiability theory of meaning in 1928; and, accordingly, he does employ the verifiability principle to attack traditional metaphysics:

> The view that [metaphysical] sentences and questions are non-cognitive was based on Wittgenstein's principle of verifiability. This principle says first, that the meaning of a sentence is given by the conditions of its verification and, second, that a sentence is meaningful if and only if it is in principle verifiable, that is, if there are possible, not necessarily actual, circumstances which, if they did occur, would definitely establish the truth of the sentence. This principle of verifiability was later replaced by the more liberal principle of confirmability. (1963a, p. 45).

And this, of course, is the basis for more straightforwardly empiricist interpretations of Carnap's underlying antimetaphysical attitude.

Two important factors militate against such interpretations, however. First, as Carnap hints at even here, the verifiability principle soon proves to be a clear – and clearly acknowledged – failure; so it can in no way explain or support Carnap's enduring antimetaphysical position.[35] Second, and more important, Carnap has available to him throughout his career more powerful and entirely independent criteria for the detection and elimination of "pseudoquestions" – namely, the purely logical devices for distinguishing between apparently well-formed statements and genuinely well-formed statements stemming from Russellian type-theory. Such purely logical criteria have nothing whatever to do with radical empiricism or any other epistemological doctrine, and Carnap appeals to them constantly in his criticisms of traditional metaphysics – in the *Aufbau*, in particular.[36] Indeed, in view of the thoroughly type-theoretic character of any constructional system, failure of constructibility in general means nothing more or less than failure to find a definite place in the type-theoretic hierarchy.

In *Logical Syntax*, this second, purely logical approach to the problem of eliminating pseudoquestions and pseudosentences becomes clearly

35 See, e.g., Carnap's (1963a, p. 38) charming account of a conversation with Einstein on this matter.

36 See Carnap (1963a, p. 45) and §§30–1 and 180 of the *Aufbau*. See also (1963a, p. 25): "Another influential idea of Wittgenstein's was the insight that many philosophical sentences, especially in traditional metaphysics, are pseudo-sentences, devoid of cognitive content. I found Wittgenstein's view on this point close to the one I had previously developed under the influence of anti-metaphysical scientists and philosophers. I had recognized that many of these sentences and questions originate in a misuse of language and a violation of logic. Under the influence of Wittgenstein, this conception was strengthened and became more definite and more radical." Note that neither radical empiricism nor the verifiability principle is explicitly mentioned here.

predominant and finds its most mature expression. However, in place of the type-theoretic hierarchy, Carnap (1934c/1937) now emphasizes the closely related distinction between object language and meta-language.[37] On the one hand, this distinction gives Carnap a clear and precise replacement for traditional metaphysics, that is, a nonempirical discipline – logical syntax – that makes possible a peculiarly philosophical vantage point from which the rest of knowledge may be surveyed:

> Metaphysical philosophy tries to go beyond the empirical scientific questions of a domain of science and to ask questions concerning the nature of the objects of the domain. These questions we hold to be pseudo-questions. The non-metaphysical logic of science, also, takes a different point of view from that of empirical science, not, however, because it assumes any metaphysical transcendency, but because it makes the language-forms themselves the objects of a new investigation. On this view, it is only possible to speak either *in* or *about* the sentences of this domain, and thus only object-sentences and syntactical sentences can be stated. (§86)

On the other hand, as the last sentence of this passage suggests, we are also given a clear and precise diagnosis of, and explanation for, the obscurities and confusions of traditional metaphysics: questions concerning the "nature" or "reality" of various entities, for example. Such questions result from attempting to employ what Carnap calls the "material mode of speech," that is, attempting to speak in both the object language and the meta-language simultaneously, as it were (§§73–81). "Philosophical sentences" in the material mode, according to Carnap, are admissible if, and only if, they are fully translatable into the "formal mode" – that is, into the meta-language of logical syntax – otherwise, they should be rejected as meaningless pseudo-sentences (§§78–81).[38]

There is, of course, no reference to empiricist epistemology or verificationism in any of this. Indeed, in §82, Carnap articulates an extremely

37 See Coffa (1991, Chapter 15) for an illuminating account of how Carnap's Russellian type-theoretic conception of logic becomes gradually transformed – largely under the influence of Tarski and Gödel – into a modern "Hilbertian" conception based on the distinction between object language and meta-language.

38 See also §75: "by this means [the diagnosis of the material mode of speech] the whole character of philosophical problems will become clearer to us. The obscurity with regard to this character is due chiefly to the deception and self-deception induced by the application of the material mode of speech. The disguise of the material mode of speech conceals the fact that the so-called problems of philosophical foundations are nothing more than questions of the logic of science concerning the sentences and sentential connections of the language of science, and also the further fact that the questions of the logic of science are formal – that is to say, syntactical – questions."

liberal and holistic epistemology, which explicitly denies that theoretical sentences can be translated into observation sentences ("protocol-sentences") and maintains that no sentence – not even a protocol-sentence – is immune from revision in the progress of science. Yet the underlying antimetaphysical attitude of Vienna is not compromised in the least:

> The syntactical problems acquire a greater significance by virtue of the anti-metaphysical attitude represented by the Vienna Circle. According to this view, the sentences of metaphysics are pseudo-sentences which on logical analysis are proved to be either empty phrases or phrases which violate the rules of syntax. Of the so-called philosophical problems, the only questions which have any meaning are those of the logic of science. To share this view is to substitute logical syntax for philosophy. (§2)[39]

The antimetaphysical dream of Vienna finally stands or falls, therefore, not with phenomenalism, radical empiricism, and the verifiability principle, but rather with the remarkable program of *Logical Syntax* itself. And this program, in turn, is best seen as a continuation and development of the earlier, and equally remarkable, program of *Der logische Aufbau der Welt.*

39 Graciela De Pierris has emphasized to me that Carnap provides a disjunctive diagnosis for the sentences of metaphysics here: they are *either* "empty phrases" *or* "phrases which violate the rules of syntax." And, whereas the second disjunct clearly accords with the present interpretation, the first could perhaps be construed as resting on verificationism after all: "empty phrases" are just those devoid of "cognitive content" – that is, unverifiable sentences. The question entirely depends, however, on what exactly Carnap means by "empty [of content]" here. According to the official definition of "content" in *Logical Syntax* (§49), being empty of content or having the "null content" is equivalent to *validity*, which, for Carnap's two official mathematical languages, is equivalent to *analyticity*. What Carnap may be saying here, then, is simply that philosophical sentences are sentences in the material mode ("pseudo-object sentences") which either can be translated into the formal mode ("syntactical sentences") or cannot be so translated. In the former case they are analytic or "empty"; in the latter they "violate the rules of syntax."

6

EPISTEMOLOGY IN THE *AUFBAU*

In *Der logische Aufbau der Welt,* Carnap inaugurates a new philosophical discipline he calls "constitutional theory [*Konstitutionstheorie*]" and presents a particular "constitutional system [*Konstitutionssystem*]" in which "[all] scientific concepts are reduced to the 'given'" (§3).[1] This particular constitutional system proceeds from an "autopsychological basis" in which "the choice of basic elements is limited to such psychological objects that belong to only one subject" (§63). More precisely, the basic elements consist of the conscious psychological objects or "experiences" of a single subject (§64). Constitutional theory also envisions other possible constitutional systems, however: notably, a constitutional system with "general-psychological basis" in which scientific concepts are reduced to the experiences of all subjects (§63), and a constitutional system with "physical basis" in which scientific concepts are reduced to the fundamental concepts of physics (§62). What is common to every such constitutional system, then, is just the circumstance that all scientific concepts are to be defined in a single system on the basis of a few fundamental concepts:

> A constitutional system does not only have the task, like other conceptual systems, of classifying concepts in various types and investigating the differences and mutual relations of these types. Rather, concepts are to be step-wise

1 Carnap (1928a/1967); references are given in the text by section number. In my quotations, I have found it necessary to deviate from the George translation at various points. In particular, although "construction theory" and "constructional system" are certainly preferable in English to "constitutional theory" and "constitutional system," the former terminology masks certain important distinctions, which will be explained below. My terminology here follows that of Sauer (1989).

> derived or "constituted" from certain basic concepts, so that a *genealogical tree of concepts* results in which every concept finds its determinate place. That such a derivation of all concepts from a few basic concepts is possible is the main thesis of constitutional theory, through which it is distinguished from most other theories of objects [*Gegendstandstheorien*]. (§1)[2]

The general discipline of constitutional theory therefore has the task of investigating all possible forms of stepwise definitional systems of concepts: all possible reductionist "system forms" (§46; compare §§59, 60).

Among the alternatives to the system form with autopsychological basis, Carnap clearly holds that the system form with physical basis is most important. This is because such a physicalistic system form "has that domain (namely the physical) as basic domain which is the only one endowed with a univocal [*eindeutige*] law-governedness of its processes" and therefore "presents the most appropriate order of concepts from the point of view of factual science [*Realwissenschaft*]" (§ 59). Indeed, Carnap was dissatisfied with the title of the *Aufbau* for precisely this reason, and he at one time envisaged a second work that was to supplement what we now know as the *Aufbau* by presenting the same kind of detailed development of a physicalistic system. This work was to be entitled *Wirklichkeitslogik* or *Der logische Aufbau der Welt*, whereas what we now know as the *Aufbau* was to be entitled *Erkenntnislogik* or *Der logische Aufbau der Erkenntnis*.[3]

Nevertheless, the *Aufbau* itself presents a detailed development of only one constitutional system, the system form with autopsychological basis, and this particular choice of system is motivated entirely by the idea of "epistemic primacy [*erkenntnismäßige Primarität*]":

> The system form that is here to be given to the outline of a constitutional system is characterized by the circumstance that it not only attempts to present the order of objects with respect to their reducibility, like every system form, but that it also attempts to present the order with respect to *epistemic primacy*. An object (or a type of objects) is called "*epistemically primary*" in relation to

2 The notion of a "theory of objects [*Gegendstandstheorie*]" refers to Meinong's *Gegendstandstheorie* (§§3, 93, 172), which investigates all objects of thought as such. As Carnap explains (§1): "The expression '*object* [*Gegenstand*]' is here used always in the widest sense, namely, for anything about which a statement can be made. Therefore we count among the objects not only things, but also properties and relationships, classes and relations, states and processes – moreover, the actual and the non-actual."

3 Carnap expresses these qualms about the title of the *Aufbau* in correspondence with Reichenbach and Schlick during 1925–7. See Coffa (1991, p. 231, n. 11). I emphasize the importance of the physicalistic alternative for constitutional theory in Chapter 5, §I (this volume).

> another – the "*epistemically secondary*" – if the latter is cognized through the mediation of the former, and therefore the cognition of the former is presupposed by the cognition of the latter. (§54)

The "intention to present through this constitutional system not only a logical-constitutional order of objects, but beyond this also their epistemological order" then motivates the choice of a system form with autopsychological basis over one with physical basis (§64; compare §59), and the same considerations motivate the choice of an autopsychological basis over a general-psychological basis (§64; compare §§58, 60). Thus the desire to have the order of logical reduction or definition reflect the order of cognition explains the particular form of constitutional system actually developed in the *Aufbau*, and it is this, moreover, that makes the *Aufbau* a work of epistemology.

What I want to explore here is the point of this "epistemic-logical" constitutional system (§1). What is the epistemological purpose of so reducing or defining all scientific concepts from an autopsychological basis, that is, from the "given"? What kind of epistemological program does the *Aufbau* represent, and in what epistemological tradition is it to be placed?

I

The answer immediately suggesting itself, especially within contemporary Anglo-American philosophy, is that the *Aufbau* belongs squarely in the tradition of modern epistemological empiricism: the tradition of the classical British empiricists, of Mach,[4] and of Russell's (1914) *Our Knowledge of the External World*.[5] The primary epistemological problem addressed by the *Aufbau* is therefore the traditional "problem of the external world": how are we justified on the basis of the immediate data of sense in the belief that there is an external world lying behind or corresponding to the immediate data of sense? How can we infer from such certain and secure data to the

4 And Mach, in turn, self-consciously places himself within the tradition of classical British empiricism. See, e.g., Mach (1886/1959, p. 368): "By studying the physiology of the senses, and by reading Herbart, I then arrived at views akin to those of Hume, though at that time I was still unacquainted with Hume himself. To this very day I cannot help regarding Berkeley and Hume as far more logically consistent thinkers than Kant."

5 Typical expressions of this conception are found in Quine and Goodman. Thus, according to Quine (1951/1963, p. 39): "[Carnap] was the first empiricist who, not content with asserting the reducibility of science to terms of immediate experience, took serious steps toward carrying out the reduction." According to Goodman (1963, p. 558): "[The *Aufbau*] belongs very much in the main tradition of modern philosophy, and carries forward a little the efforts of the British Empiricists of the 18th Century."

apparently much less certain claims of science and common sense? The epistemological point of the *Aufbau* is to develop a traditional phenomenalist or reductionist solution to this problem: the external world does not lie behind or correspond to the immediate data of sense at all; rather, it is nothing but a complex logical construction out of such immediate data. Our claims about the external world are, in the end, complex claims about the immediate data of sense and hence are thereby justifiable in principle. What then distinguishes the *Aufbau* within the empiricist tradition is simply the greater detail and rigor with which it attempts to carry out this phenomenalist program.[6]

This conception of the epistemological point of the *Aufbau* is certainly a very natural one. Carnap takes as his motto for the book Russell's "supreme maxim in scientific philosophizing": "Whenever possible, logical constructions are to be substituted for inferred entities." References to Mach and especially to Russell are found frequently throughout the text. In his "Intellectual Autobiography," Carnap explicitly names Russell's *Our Knowledge of the External World* as the central stimulus and inspiration for his writing of the *Aufbau*;[7] and he then articulates the above epistemological conception in the most explicit terms:

> Under the influence of some philosophers, especially Mach and Russell, I regarded in the *Logischer Aufbau* a phenomenalistic language as the best for a philosophical analysis of knowledge. I believed that the task of philosophy consists in reducing all knowledge to a basis of certainty. Since the most certain knowledge is that of the immediately given, whereas knowledge of material things is derivative and less certain, it seemed that the philosopher must employ a language which uses sense-data as a basis. (1963a, p. 50)

> [The Vienna Circle] assumed that there was a certain rock bottom of knowledge, the knowledge of the immediately given, which was indubitable. Every other kind of knowledge was supported by this basis and therefore likewise decidable with certainty. This was the picture which I had given in the

6 Quine (1969, p. 74) puts the point thusly: "To account for the external world as a logical construct of sense data – such, in Russell's terms, was the program. It was Carnap, in his *Der logische Aufbau der Welt* of 1928, who came nearest to executing it."

7 See Carnap (1963, pp. 13, 16). In Carnap's personal copy of *Our Knowledge of the External World*, at the end of the third chapter, where Russell speculates that his construction "can be obtained from more slender materials by the logical methods of which we shall have an example in the definitions of points, instants, and particles," Carnap wrote in the margin: "This narrowing and deepening of *the fundamental* postulates is my task!" See the Introduction to Creath (1990, p. 24). Carnap's copy of Russell's book can be found in the Archives for Scientific Philosophy (ASP), University of Pittsburgh Libraries. Subsequent references are to file folder numbers.

> *Logischer Aufbau*; it was supported by the influence of Mach's doctrine of the sensations as the elements of knowledge, by Russell's logical atomism, and finally by Wittgenstein's thesis that all propositions are truth-functions of the elementary propositions. (p. 57)[8]

It would be difficult indeed to find a clearer statement anywhere of the assumptions and goals of phenomenalistic foundationalism.

Yet when we turn to the text of the *Aufbau* itself, such an epistemological conception is hardly in evidence. First, Carnap, as a matter of fact, devotes very little space or energy to the problem of the external world. Rather, most of his effort is devoted to a technical elaboration of the procedure of "quasi-analysis" by which specific sensory qualities such as colors are defined from originally undifferentiated momentary cross sections of the "stream of experience," that is, from "elementary experiences" (§§67–93, 104). This construction, which takes place entirely within the domain of the autopsychological, is then the only part of Carnap's "Outline of a Constitutional System" actually to be presented in complete logical detail (§§108–21). The constitution of physical objects – comprising the "visual things," "my body," the "tactual-visual things," "the perceptual things," the "world of physics," and finally the "biological objects" including "human beings"–is then presented only briefly and sketchily (§§124–37). The key step in the constitution of the external world is actually presented in a single section, "The ascribing of colors to world-points" (§126): after constructing space-time as a purely mathematical object out of *n*-tuples of real numbers (§125), Carnap embeds the previously defined visual fields (§117) of our subject into this space-time and projects colored points of the visual field onto colored points external to the subject in space in such a way that principles of continuity and constancy are satisfied.

In no domain beyond that of the autopsychological does Carnap make any attempt whatever to present genuine logical definitions of the constituted objects. In particular, he makes no attempt to show that his principles for ascribing colors to world points can be turned into an explicit definition of the visual things (§128); and this, of course, is why it is now standardly thought that Carnap's "construction of the external world" is a failure.[9]

8 I am indebted to Silvana Gambardella for emphasizing the importance of the second passage to me. This passage also is discussed by Creath (1982).

9 See Quine (1951/1963, p. 40; 1969, pp. 76–7). Carnap himself makes the same point in the Preface to the second edition of the *Aufbau*: "Without myself being clearly conscious of it I in fact already went beyond the limits of explicit definition in the constitution of the physical world. For example, for the coordination of colors to space-time points (§127f.) only general principles were stated but not univocal [*eindeutige*] operational prescriptions."

What I want to emphasize here, however, is that Carnap in fact devotes very little attention to this problem and prefers instead to concentrate on other matters, namely, the construction of the entire domain of the autopsychological from a single primitive relation (remembrance of part-similarity in some arbitrary respect, §78) holding between unanalyzable (§68) elementary experiences. The idea that Carnap is presenting here a more detailed and rigorous solution to the problem of the external world than had Russell is therefore, in this respect at least, seriously at odds with the text.[10]

In the second place, Carnap nowhere employs the traditional epistemological vocabulary of "certainty," "justification," "doubt," and so on in the *Aufbau*.[11] He nowhere says that knowledge of autopsychological objects is more certain or more secure than knowledge of physical objects, and the distinction between "hard data" and "soft data" central to Russell's motivation for his construction of the external world is entirely foreign to the *Aufbau*. Carnap appeals at all levels of his constitutional system to "the particular results of the factual sciences [*Realwissenschaften*]" in guiding his specific methods of constitution (§122). Indeed, this is particularly true at the autopsychological level, where the choice of unanalyzable

Technically speaking, the coordination of colors (later, sensory qualities in general and then, physical state magnitudes) to space-time points will fail to yield an explicit definition if in the hierarchy of types the procedure nowhere closes off at a definite rank. Since Carnap himself describes the procedure as subject to a complex process of continual revision (§135), it appears very likely that this last eventuality is indeed realized. For further discussion, see the Postscript to this chapter.

10 When Carnap speaks of a "narrowing and deepening of [Russell's] *fundamental* postulates" in the margin of his copy of *Our Knowledge of the External World* (see footnote 7, this chapter), it is quite possible that he has in mind not a more rigorous construction of the external world, but rather precisely the above construction of the purely autopsychological domain. Indeed, in 1922, shortly after having first read Russell's book in 1921, Carnap wrote a manuscript, "Vom Chaos zur Wirklichkeit," largely devoted to a primitive version of this construction; at the top of the manuscript, Carnap later wrote "this is the germ for the constitutional theory of the 'Logical Aufbau'" (ASP, RC 081-05-01). Moreover, Carnap (1924) discusses the relationship between the "primary world" of immediately given sensations and the "secondary world" of physical objects; Carnap there deliberately leaves to one side the question of constructing the primary world itself from an "original chaos" as "a question of epistemology" (1924, p. 108). Finally, in the passage from *Our Knowledge of the External World* on which Carnap is commenting in his marginal note, Russell refers to the "logical methods" of the fourth chapter in which points, instants, and so on are constructed as equivalence classes; Carnap refers explicitly to this chapter of Russell in developing his own technique of "quasi-analysis" in the *Aufbau* (§73).

11 Carnap does employ this vocabulary in Carnap (1928b/1967). However, as Carnap makes clear in his Preface to the English (second) edition of the *Aufbau*, Carnap (1928b) belongs to a different period of thought, after he had moved to Vienna in 1926, and accordingly "shows a stronger influence of the Vienna discussions and Wittgenstein's book." By contrast, the *Aufbau* was written largely in the years 1922–5: see Carnap (1963a, pp. 16, 19–20).

elementary experiences as basic elements is based principally on the empirical findings of Gestalt psychology (§67). Carnap's aim, accordingly, is to demonstrate "[t]he *translatability of all scientific assertions into assertions within a constitutional system*" (§122) in such a way that "the actual process of cognition" is "*rationally reconstructed*" (§143; compare §100). In this way, the constitutional system presents a rational reconstruction of the order in which objects of various domains are, in fact, cognized, but there is no suggestion at all that objects of different levels differ in certainty or security of epistemic value. On the contrary, as parts of a unified presentation of the results of the empirical sciences, all objects of the constitutional system necessarily have the same (tentative and empirical) epistemic value.[12]

The order of cognition reflected in the constitutional system is, as noted above, the order of epistemic primacy (§54): lower-level objects are epistemically primary relative to higher-level objects, and this means that "the latter [are] cognized through the mediation of the former, and therefore the cognition of the former is presupposed by the cognition of the latter." Note the blandness and neutrality of this characterization: there is no implication, in particular, of different degrees of certainty or security.[13] A similar blandness characterizes the discussion of the "method of indicators" that actually puts the order of epistemic primacy into effect. Thus an "indicator" of a state of affairs is "such a [sufficient] condition by which the state of affairs is also customarily cognized, which is therefore usually cognized *before* the state of affairs" (§49). As examples, Carnap then gives the barometer as an indicator for air pressure and "x is an animal which carries a number of rattles at the end of its body" as an indicator for "x is a rattlesnake."[14] It is hard

12 See especially §106: "The content [of the constitutional system] depends on the contentful results of the factual sciences – indeed, the lower levels in particular depend on the results of the phenomenology of perception and psychology. Since the results of these sciences are themselves still controversial, the thoroughgoing contentful correctness of their translation into the language of a constitutional system cannot be guaranteed."

13 It is clear, moreover, that Carnap takes the order of epistemic primacy to be entirely uncontroversial. He assumes without comment, for example, that all epistemological "tendencies" will agree that "all cognition traces back finally to my experiences, which are set into relation, connected, and worked up; thus cognition can attain in a logical progress to the various structures of my consciousness, then to the physical objects, further with their help to the structures of consciousness of other subjects and thus to the heteropsychological, and through the mediation of the heteropsychological to the cultural objects" (§178).

14 As Howard Stein has pointed out to me, the original uses the example of the "hooded cobra [*Brillenschlage*]," which "is an animal that behind its head bears the figure of a bent pair of spectacles [*Brille*]." George's substitution nicely captures the duplication of words here.

to see how any serious work in traditional justificational or foundationalist epistemology can possibly be done here.[15]

Nevertheless, it is instructive to trace out in more detail how Carnap's method of indicators in fact applies to the traditional problem of the external world. As noted above, the crucial step in the constitution of the external world is taken in §126, where Carnap projects colored spots in the subject's (two-dimensional) visual field onto external points in (three-dimensional) space. The indicator of a colored point on the surface of a physical object is therefore a similarly colored spot in my visual field. Does it follow that there is such a colored surface whenever I sense a similarly colored spot in my visual field? Of course not; for the coordination of colored physical points to colored spots in my visual field is further regulated and controlled by principles of continuity and constancy – and then supplemented and corrected by analogy (§135), the general laws of physics (§136), and the reports of other persons (§144). Intuitively, such supplementation and correction may show that the original colored spot in my visual field was an "illusion" and thus proceeded from another "cause" than a real colored surface.

Now such phenomena of "illusion" and "multiple causation" are, of course, taken traditionally to show that the coordination of physical objects to sense data is not unique and to argue therefrom to the conclusion that physical objects *cannot* be defined in terms of sense data. Indeed, Carnap (1924) presents just such an argument (see footnote 10, this chapter). Carnap's principal claim there is that the two-dimensional "primary world" of immediately given sensations exhibits no "determining laws [*determinierende Gesetze*]" by which processes are "univocally determined [*eindeutig bestimmt*]." Only the three-dimensional "secondary world" of physics exhibits this determination, and thus the two "fictions" of three-dimensional space and thoroughgoing causal determination are bound inextricably together. In the course of his argument, Carnap consider the following objection: the primary world is subject to determining laws after all, for all we have to do is substitute into the laws of physics the sensory qualities that are univocally

15 There is one passage in the discussion of indicators (§49) where Carnap uses more epistemically loaded vocabulary. He asserts that "for every scientific state of affairs there is in principle a *simultaneously infallible and never absent indicator* [*zugleich untrügliches und nie fehlendes Kennzeichen*]." However, he immediately explains the emphasized phrase thusly "i.e., an indicator that is always then present, but also only then present, when the state of affairs is present." The rattles and the rattlesnake then are introduced precisely to illustrate *this* relation. I am indebted to Alan Richardson for emphasizing the importance of this passage to me.

[*eindeutig*] correlated with physical state-magnitudes, and physical laws then translate into laws governing sensations. Carnap replies precisely that the coordination between the secondary world and the primary world is many-one rather than one-one [*nicht eineindeutig, sondern mehreindeutig*]; therefore, it is not possible to translate physical laws into laws governing sensations (1924, p. 126).

Interestingly enough, Carnap presents parallel considerations in the *Aufbau*. Here the constitution of the world of physics takes place via the "physical-qualitative coordination" between the perceptual world (itself a coordination first of colors and then of other sensory qualities to external spatial points) and the purely quantitative, numerical state-magnitudes of physics. However, the physical-qualitative coordination is "a one-many coordination [*einmehrdeutige Zuordnung*] between qualities and state-magnitudes," which does not correlate a univocal [*eindeutig*] state-magnitude with a given (physical) sensory quality (§136). How then is it possible to constitute physical state-magnitudes in terms of such sensory qualities? How can we univocally translate statements about the former into statements about the latter? More generally, since the perceptual world is itself supplemented and corrected in light of the laws of physics, how is it possible to constitute the external world from sensations at all? And how is it then possible, as Carnap claims in §179, to translate all statements of science into "statements about the basic objects, namely, about relations between elementary experiences"?

Section 136 of the *Aufbau* refers us to Carnap (1923) for more details on the physical-qualitative coordination. Although Carnap repeats the claim that the coordination between "phenomenal facts" and corresponding state-magnitudes is only univocal in the direction from the latter to the former, he there outlines a procedure for nonetheless approximating to a univocal assignment of physical state-magnitudes by focusing on a small neighborhood of a given phenomenally characterized space-time point and working back and forth using the laws of physics (1923, pp. 102–3). The crucial point is that the laws of physics, together with a univocal determination of phenomenal qualities from physical state-magnitudes, provide a methodological procedure for narrowing down the ambiguity in the assignment of physical state-magnitudes: in principle, a univocal assignment is thereby constructed after all. It appears, then, that, in §136 of the *Aufbau*, Carnap intends to achieve a univocal constitution of the world of physics by just such a methodological procedure.

But what is the epistemological status of the laws of physics and the resulting methodological procedure? The answer of Carnap (1923) is perfectly clear: they are the result of conventional choice or "stipulation [*Festsetzung*]" subject to "the principle of simplicity." And it is clear from §136 that this is the position of the *Aufbau* as well: the world of physics is univocally determined via the physical-qualitative coordination *plus* conventional stipulations.[16] More generally, the same result holds for the constitution of the visual things in §§126–8: the assignment of colors to world points becomes univocal only in the context of the methodological principles of continuity and constancy (together with the further supplementations and corrections noted above); and, although Carnap never discusses the matter explicitly, it is clear that these principles, too, are conventions or stipulations.[17] In the end, therefore, the entire constitution of the external world is determined from sensory data on the basis of a complicated system of physical and methodological conventions or stipulations that are intended to do nothing more or less than encode the actual (although largely unconscious) rules that science follows in constructing its picture of the physical world. Only such a complicated system of conventions enables us – in principle, Carnap hopes – to translate all statements of science into statements about elementary experiences.[18]

16 Carnap there says that the constitution of the world of physics, "neglecting the law-governedness which is to be introduced therein," is essentially determined through the physical-qualitative coordination. He alludes to the choice of a system of physics according to "the principle of simplicity" and explicitly refers to Carnap (1923) for further discussion of this choice. Compare also the list of possible systems of physics in §62 of the *Aufbau* with that in Carnap (1923, §III).

17 In earlier characterizing the analogous "general rules of constitution," Carnap writes (§103): "These general rules can be designated as *a priori* rules, in so far as the constitution and the cognition of objects logically rests on them. . . . The rules are not to be designated as "a priori cognition," however, for they present us not with cognitions, but rather *stipulations* [*Festsetzungen*]. In the actual cognitive-process these stipulations occur unconsciously. Even in scientific procedures they are seldom made consciously and explicitly."

18 Carnap continues to hold, as noted earlier, that only the world of physics exhibits "a univocal [*eindeutige*] law-governedness of its processes." Since he now holds that statements about the world of physics *are* translatable into statements about the immediate data of sense, how is this consistent with the argument of Carnap (1924), according to which the "primary world" fails to be law-governed precisely because the "secondary world" cannot be univocally translated into sensory terms? The answer is that Carnap is now working within Russell's theory of types. Although the physical world is translatable into sensory terms, the domain of the physical and the domain of the autopsychological still belong to entirely different "object-spheres" or logical types (1928b/1967, §§30–1): all higher level objects are classes of classes . . . of classes of elementary experiences. That the domain of the physical is univocally law-governed therefore does not imply that the domain of the autopsychological is as well. (In terms of footnote 9, this chapter, however, translatability will nonetheless fail if the complex methodological procedure in question terminates at no definite rank in the hierarchy of types.)

It follows that Carnap's ultimate solution to the problem of the external world – in so far as such a solution is present at all in the *Aufbau* – is very far indeed from traditional empiricism and phenomenalistic foundationalism.[19] For the problem of the external world is finally solved, not simply in virtue of a purely sensory translation, but rather by the idea that the methodological procedures and assumptions actually deployed in developing our claims about the physical world are to be characterized as conventions or stipulations rather than as cognitions; accordingly, a demand for *their* "justification" is entirely inappropriate. In this way, rather than presenting a traditional empiricist or phenomenalist account of our knowledge of the external world, the *Aufbau* instead anticipates Carnap's (1950a) later strategy of "Empiricism, Semantics, and Ontology": the question of the reality of the external world dissolves into the "external question" of whether or not to accept and use the forms of expression of the "thing language." Such an "external question" is not, of course, subject to rational dispute at all, but only to purely pragmatic considerations of convenience.[20]

II

The preceding considerations suggest that the customary assimilation of the epistemological project of the *Aufbau* to that of the empiricist philosophical tradition has perhaps been too hasty. In §75, where Carnap first introduces the basic relation of his "logical-epistemological" constitutional system, he explicitly aligns his project with a rather different tradition:

19 The nonempiricist character of Carnap's strategy stands out particularly clearly in Carnap (1923). Carnap there begins as follows: "After a long time during which the question of the sources of physical cognition has been violently contested, it may perhaps already be said today that pure empiricism has lost its dominance. That the construction of physics cannot be based on experimental results alone, but must also apply non-empirical [*nichterfahrungsmäßige*] principles has indeed been already proclaimed for a long time by philosophy" (1923, p. 90). He then expounds the "conventionalism" of Poincaré and Dingler (1923, §I). Finally, he characterizes the various axiomatic systems of physics as consisting of "*synthetic a priori propositions,* however not exactly in the Kantian transcendental-critical sense. For this would mean that they express necessary conditions of the objects of experience, themselves conditioned by the forms of intuition and of thought. But then there could be only *one* possible form for the content of [such a system]. In reality, however, its construction is left in many ways to our choice" (1923, p. 97).

20 See Carnap (1950a/1956, §2). "Internal questions" *within* the "thing language," by contrast, "are to be answered by empirical investigations. Results of observations are evaluated according to certain rules as confirming or disconfirming evidence for possible answers. (This evaluation is usually carried out, of course, as a matter of habit rather than a deliberate, rational procedure. But it is possible, in a rational reconstruction, to lay down explicit rules for the evaluation. This is one of the main tasks of a pure, as distinguished from a psychological, epistemology.)"

Cassirer ([Substanzbegr.] 292ff.) has shown that a science having the goal of determining the individual through contexts of laws [*Gesetzseszusammenhänge*] without its individuality being lost must apply, not class ("species") concepts, but rather *relational concepts*; for the latter can lead to the formation of series and thereby to the establishing of order-systems. It hereby also results that relations are necessary as first posits, since one can in fact easily make the transition from relations to classes, whereas the contrary procedure is only possible in a very limited measure.

The merit of having discovered the necessary basis of the constitutional system thereby belongs to two entirely different, and often mutually hostile, philosophical tendencies. *Positivism* has stressed that the sole *material* for cognition lies in the undigested [*unverarbeitet*] experiential *given*; here is to be sought the *basic elements* of the constitutional system. *Transcendental idealism*, however, especially the neo-Kantian tendency (Rickert, Cassirer, Bauch), has rightly emphasized that these elements do not suffice; *order-posits* [*Ordnungssetzungen*] must be added, our "basic relations."[21]

Carnap here associates his project with recent developments within the tradition of neo-Kantian epistemology: in particular, with Ernst Cassirer's (1910) *Substanzbegriff und Funktionsbegriff*, Heinrich Rickert's (1892) *Der Gegenstand der Erkenntnis*, and Bruno Bauch's (1923) *Wahrheit, Wert und Wirklichkeit*.[22]

This neo-Kantian tradition approaches epistemology in different terms, and from a different point of view, than does the empiricist epistemological tradition more familiar within contemporary analytic philosophy. The primary problem does not involve the justification of our beliefs, the refutation of philosophical skepticism, or the relative degrees of certainty and epistemic value of beliefs in various different categories. Instead, such

21 The importance of this passage, together with other passages referring to neo-Kantianism considered later, has been rightly emphasized by Sauer. See the article referred to in footnote 1, this chapter, and also Sauer (1985, 1987). My discussion throughout this section proceeds in very substantial agreement with Sauer; my disagreements with him will emerge in the following section. For a somewhat different perspective, see also Richardson (1992).

22 Cassirer was the most important contemporary representative of the Marburg School of neo-Kantianism established by Hermann Cohen; of all the neo-Kantians, he is clearly the one who is most attentive to recent developments in physics, mathematics, and logic. Rickert and Bauch belonged to the Southwest School of neo-Kantianism founded by Wilhelm Windelband. Rickert was Bauch's dissertation advisor at Freiburg (and Bauch dedicates his book to Rickert); Bauch, in turn, was Carnap's dissertation advisor at Jena – see Carnap (1963a, pp. 4, 11–12). Some discussion of Bauch in relation to both Frege (his colleague at Jena) and Carnap can be found in Sluga (1980); the discussion of Carnap in particular occurs on pp. 178–81, although Sluga there incorrectly asserts that both Bauch and Frege directed Carnap's dissertation. It should be especially emphasized, finally, that when I speak of "neo-Kantian epistemology" in what follows, I am limiting myself specifically to Cassirer, Rickert, and Bauch, who were undoubtedly the most important neo-Kantian influences on Carnap himself.

neo-Kantian philosophers occupy themselves with what they take to be the prior problem of how "objective judgments" are possible in the first place: what makes such things as judgments – which are essentially capable of either truth or falsity, justification or disconfirmation – possible? How does it come about that our thought, which initially appears to be confined to merely subjective representations or ideas intrinsically possessing neither truth nor falsity, nonetheless acquires objective meaning or "relation to an object" so that questions of truth and falsity (and thus questions of epistemic justification) then apply?

Neo-Kantian epistemology begins with the conviction that neither "strict empiricism" nor "metaphysical realism" can provide a satisfactory solution to these problems. Strict empiricism is unsatisfactory because no such thing as an objective judgment can possibly be found among the essentially private, fleeting, and at best vaguely differentiated immediate data of sense; such data merely occur, but are neither true nor false, neither justifiable nor unjustifiable. Metaphysical realism is also unsatisfactory, however, because it attempts to base objectivity on the relation of sensory data to a "transcendent" object existing somehow behind the data, and it therefore cannot explain how access to objects and thus objective judgment is possible for us. Since it is clear, in any case, that our cognition must start from the immediate data of sense, metaphysical realism simply creates an unbridgeable gulf between thought and reality in virtue of which objective judgments are just as impossible for us as they are on a strictly empiricist or "positivist" conception. The problem, then, is to construct a new conception of "relation to an object" and thus of "reality" that shows how we can proceed from private, subjective sense impressions to truly objective judgments without positing transcendent objects existing behind our sensory data, that is, without positing *Dinge an sich.* In this way, the "problem of cognition [*Erkenntnisproblem*]" is intimately connected, for the neo-Kantians, with the "problem of reality [*Wirklichkeitsproblem*]."

The basic idea of the neo-Kantian solution is expressed in the paradoxical-sounding formula: the "real" or "actual" is made possible by the "unreal" or "non-actual." The "unreal" or "non-actual" is the realm of purely objective, timelessly valid laws of logic and mathematics: the realm of "necessities of thought [*Denknotwendigkeit*]." Following Lotze, we carefully distinguish between the realm of what "exists" or is "real" and the realm of what "holds [*gelten*]" or is "valid [*gültig*]."[23] In particular, the realm of validity is not

23 See Lotze (1874/1884, §§316–20). Lotze actually distinguishes three realms: things, which have "being [*Sein*]," events, which "happen [*geschehen*]," and propositions or relations

constituted by existent entities or *Dinge an sich* located outside the sphere of our thought but rather by normative and objective (i.e., intersubjectively valid) rules or laws regulating or governing our thought. Such objective rules are exhibited, first and foremost, in pure mathematics and pure logic, but this in no way exhausts their epistemological function. On the contrary, their peculiarly epistemological function is precisely to transform the immediate data of sense by means of mathematics and logic into the objects of mathematical natural science: we thereby create or generate the world of reality or actuality. Sensation thus acquires "relation to an object" if, in Bauch's words, "it is arranged and adjusted in a context, precisely the context of objective necessities of thought or laws of validity."[24]

Cassirer (1910/1923, Chapter IV, §II) illustrates the "peculiar interweaving of 'actual' and 'non-actual' elements, on which all natural-scientific theories rests" in a particularly striking way in his discussion of "ideal limiting structures" such as moving point-particles and the like. He diagnoses the "skeptical" and "empiricist" arguments of P. du Bois-Reyond, which reject such limiting structures on account of their unobservability, as resting on an incorrect conception of "the relation of *concept* to *existence*, of *idea* to *reality*." In fact, du Bois-Reymond's arguments present us with a false dichotomy:

> We must choose between these two world-views: either, with *empiricism*, we only posit as present that which can be individually exhibited in the actual representation, or, rather, with *idealism*, we assert the existence of structures that form the conclusion in thought of determinate series of representations but are never themselves immediately given. (1910/1923, p. 123)

The dichotomy is false because it ignores precisely the circumstance that empirical actuality is generated by *idealizing* what is sensibly given in terms of the eternally valid laws of logic and mathematics:

> The aggregate of sensible things must be related to a system of necessary concepts and laws and be brought together to unity in this relation. But this process of thought certainly requires more than the mere combination and

between things, which have "validity [*Geltung*]." Rickert (1892, pp. 150–1) refers to Lotze (along with Brentano, Sigwart, Bergmann, and Windelband) as a forerunner of his conception of objective judgment. Bauch (1923, pp. 3–4, 240–1) refers to Lotze (along with Rickert) in this connection also. As both Rickert (1892, p. 239n) and Bauch (1923, pp. 11–12) point out, there is a close connection between these ideas and Meinong's *Gegendstandstheorie*: compare footnote 2, this chapter.

24 Bauch (1923, p. 133): "*sie eingeordnet und eingestellt ist in einen Zusammenhang, eben den Zusammenhang objektiv-denknotwendiger Geltungsgesetzlichkeit.*"

> deformation of constituents of representations; it presupposes an independent and constructive activity, as is manifested most clearly in the creation of limiting structures. And *this* form of idealization must be also granted by the "empiricist," for without it the perceptual world would be not only a mosaic, but a true chaos. (p. 128)

No awkward questions concerning the "subsistence" of ideal limiting structures behind the phenomena will arise here (p. 129): "For the subsistence [*Bestand*] of the ideal, which can alone be critically asserted and represented, asserts no more than the objective logical necessity of the idealization."

Cassirer concludes that we can proceed from private, subjective sense impressions to truly objective judgments, as desired, in a purely logical-epistemological fashion. "Transcendent-metaphysical" concepts are in no way required:

> Certainly the *metaphysical* concept of "transcendence" lies wholly outside this progress from the mere process of sense-impressions to determinate "objective" assertions. The transformation that takes place here, and which the natural-scientific concept first produces and makes possible, provides the sense-data with a new *form of being* [*Seinsform*] only in so far as they impress upon them a new *form of cognition* [*Erkenntnisform*]. (p. 242)

And, as we know, the required new form of cognition is impressed upon the data of sense by their being "brought together to unity" in a "context" provided by logically necessary objective laws of thought. In this sense, neo-Kantian epistemology endorses Russell's "supreme maxim in scientific philosophizing" as enthusiastically as does empiricist-phenomenalist epistemology.

This last point is centrally important, for it makes it clear that there can be other philosophical motivations for proceeding from immediate sense experience via "logical construction" to the totality of our scientific knowledge than the motivations of traditional empiricism and phenomenalistic foundationalism. Indeed, as we have seen, the neo-Kantian motivations for undertaking such an epistemological project are, in an important sense, precisely the reverse of those of traditional empiricism. In the empiricist tradition the immediate data of sense constitute the paradigm of knowledge and certainty: no epistemological doubts can possibly arise here. The point of a foundationalist "logical construction" on the basis of such data is then to transfer, as far as possible, the epistemic value and certainty of the immediate data of sense to the rest of our scientific knowledge. In the neo-Kantian tradition, by contrast, the immediate data of sense do not, by themselves, constitute objective knowledge at all. Such data are essentially subjective, private, fleeting, and imprecise; so no objective knowledge can possibly be found here. The point of proceeding from the data of sense via "logical construction" to our scientific knowledge is not, therefore, to transfer the epistemic status of the former to the latter, but rather to embed the data of sense in an objective logico-mathematical structure so that they themselves first become objective.[25] We might express the essential difference thus: in the first tradition, *certainty* flows, as it were, from the bottom up, whereas in the second tradition, *objectivity* flows from the top down.

Now which type of philosophical motivation is most evident in the *Aufbau*? We saw earlier in §I of this chapter that the motivations of traditional empiricism appear to be hardly in evidence at all. Motivations closely akin to those of the neo-Kantians, however, are very clearly expressed. Indeed, the entire point of Carnap's technique of "quasi-analysis" is to make possible what he calls "purely structural definite descriptions" of the various sense qualities and sense modalities: definitions that individuate the various types of sensory objects in purely formal-logical terms making no reference whatever to their intrinsic phenomenal qualities. The visual field, for example, is the unique sense modality having exactly five dimensions – two of spatial location and three of hue, saturation, and brightness (see §§86, 88–91) – and this purely logical characterization suffices to distinguish the visual field from every other sense modality wholly independently of intrinsic sensory "qualitativeness" (see especially §91). Moreover, the point of this kind of characterization is precisely to transform apparently private and subjective entities into intersubjective, objective entities:

> *every scientific statement can in principle be so transformed that it is only a structural statement.* But this transformation is not only possible, but required. For science wants to speak about the objective; however, everything that does not belong to structure but to the material, everything that is ostended concretely, is in the end subjective.
>
> From the point of view of constitutional theory this state of affairs is to be expressed in the following way. The series of experiences is different for each subject. If we aim, in spite of this, at agreement in the names given for the objects [*Gebilde*] constituted on the basis of the experiences, then this cannot occur through reference to the completely diverging material but only through the formal indicators of the object-structures [*Gebildestrukturen*]. (§ 16)

25 This theme is particularly clear in Bauch, who views "the inclusion of sensation in the context of necessities of thought or laws of validity as presupposition of sensation itself [*die Einbezogenheit der Empfindung in den Zusammenhang denknotwendiger Geltungsgesetze als Voraussetzung der Empfindung selbst*]" (1923, p. 133).

In this way, the constitutional system demonstrates that objective knowledge is possible *despite* its necessary origin in purely subjective experience.[26]

The epistemological significance of purely structural definite descriptions stands out most clearly, perhaps, in Carnap's discussion of the autopsychological basis and "methodological solipsism" (§§64–6). Carnap begins by pointing out that philosophical resistance to the idea of an autopsychological basis stems largely from the conviction that it is impossible to proceed from such a basis to an intersubjective world common to all subjects (§64). He then argues (§65) that "the given is subjectless" and does not asymmetrically single out one experiencing subject from all others:

> The expressions "autopsychological basis" and "methodological solipsism" are not to be so interpreted as if we intended initially to separate the "*ipse*," the "I," from the other subjects, or as if one of the empirical subjects were singled out and declared to be the epistemological subject. Initially, there can be no question of either other subjects or the I. Both are first constituted – and indeed together – at a later stage.

Finally, §66, "The problem of objectivity within an autopsychological basis," refers back to the discussion of purely structural definite descriptions in §§15–16 and reemphasizes their importance in the constitution of objectivity:

> Only on the basis of this recognition, that *science is essentially structural-science* and that *therefore there is a way to constitute the objective proceeding from the individual stream of experience*, is the system form with autopsychological basis admissible.

Purely structural definite descriptions, and such definite descriptions alone, make the objectivity of cognition possible.

This solution to "the problem of objectivity within an autopsychological basis" should be compared with the discussion of the same problem by Mach (1886/1959, Chapter I, §12), a discussion to which Carnap refers twice in the course of his own account of the problem (§§64, 65). For Mach, the problem of objectivity is solved solely on the basis of the circumstance that (in Carnap's terms) the given is subjectless. Mach's "elements" or sensory contents do not come initially attached to a self: the self, bodies, and other selves

26 See §2: "Although the subjective starting point of all cognition lies in the contents of experience and their interconnections, it is still possible, as the construction of the constitutional system is to show, to arrive at an intersubjective, *objective world*, which is conceptually comprehensible – and, indeed, as an identical world for all subjects." Burton Dreben has rightly emphasized to me that Carnap is here indebted to Russell's (1919, Chapter 4) prior delineation of a closely related connection between logical structure and objectivity, as Carnap himself points out in §16.

are only constructed subsequently in terms of differing organizations of the elementary given. Since the self, just as much as bodies and other selves, is "an ideal mental-economical unity, not a real unity," the elementary sensory contents are not confined to a particular individual, and intersubjectivity is therefore not a problem. For Carnap, by contrast, the circumstance that the given is thus subjectless in no way suffices to solve the problem of objectivity. On the contrary, this problem is solved only by self-consciously taking the additional step of defining all concepts of science purely structurally and by thereby embedding sensory contents themselves within a formal-logical "context," namely, the theory of relations and theory of types provided by *Principia Mathematica*.[27] It is this that decisively separates the epistemological project of the *Aufbau* from that of Machian "positivism" and, as Carnap himself explicitly notes in §75, aligns his project rather with the "transcendental idealism" of the neo-Kantians.[28]

Carnap explains the conception common to neo-Kantian "transcendental idealism" and his own constitutional system in striking terms in the course of his discussion of the problem of reality [*Wirklichkeitsproblem*] in §177:

> Constitutional theory and *transcendental idealism* agree in representing the following position: all objects of cognition are constituted (in idealistic language, are "generated in thought"); and, indeed, the constituted objects are only objects of cognition as logical forms constructed in a determinate way. This holds ultimately also for the basic elements of the constitutional system. They are,

27 An additional important respect in which Carnap's procedure is essentially different from that of both Mach and Russell is that Carnap introduces space(-time) as a primitive, purely mathematical object (§122) and makes no attempt to construct space (as both Mach and Russell do) from sensory data. And this, moreover, explains why Carnap can dispense with both the sense data of other people and with *sensibilia*: since all points of space are already given purely mathematically, there is no need to construct "points of view" that neither I nor possibly anyone ever takes up (see §§3, 64, 124, 140). Yet it is ironic that the one respect in which Carnap thus "improves" on Russell's construction of the external world in fact distances the *Aufbau* even further from a strictly empiricist epistemology (here compare footnotes 7, 10, and 19, this chapter).

28 The first paragraph of §75, with its mention of the problem of "individuality," refers to a dispute between Rickert and Cassirer that is most relevant here. Rickert (1902/1986) argues that, although the concepts of natural science indeed confer objectivity on what is sensibly given, they also inevitably result in a loss of individuality and immediate concreteness. The latter can only be comprehended by the quite different method of *historical* concept formation. Cassirer (1910/1923, Chapter IV, §IX) then argues that logico-mathematical *relational* concepts are not vulnerable to Rickert's complaint; and it is this discussion of Cassirer's to which Carnap is referring in the first paragraph of §75. In §12, Carnap points out, again referring to this discussion of Cassirer's (and also to Rickert, Windelband, and Dilthey), that the desired "logic of individuality" can be attained precisely by the method of structural description.

> to be sure, taken as basis as unanalyzed unities, but they are then furnished with various properties and analyzed into (quasi-) constituents (§116); first hereby, and thus also first as constituted objects, do they become objects of cognition properly speaking – and, indeed, objects of psychology.[29]

Since §116 (compare §93) presents the actual constitution of *sensations*, defined via a purely structural definite description containing only the basic relation itself as nonlogical primitive, the neo-Kantian conception, which views "the inclusion of sensation in the context of necessities of thought or laws of validity as presupposition of sensation itself" (footnote 25, this chapter), could hardly find clearer or more precise expression.

III

In so far as traditional epistemological motivations are present at all in the *Aufbau*, those of the neo-Kantian tradition are therefore much more explicitly in evidence than those of the empiricist tradition.[30] Nevertheless, it would be just as mistaken to assimilate the epistemology of the *Aufbau* to neo-Kantianism as it is to assimilate it to empiricism. Indeed, in the two texts where neo-Kantian motivations are most clearly expressed – §75 and §177 – Carnap also explicitly underscores his agreement with "positivism" and phenomenalistic empiricism. In §177 the latter position is subsumed under the rubric of "subjective idealism":

> Constitutional theory and *subjective idealism* agree in that all statements about objects of cognition can be in principle transformed into statements about structural interconnections [*Strukturzusammenhänge*] of the given (with retention of logical value, see §50). With *solipsism* constitutional theory shares the conception that this given consists of my experiences.

And this agreement between the constitutional system and (Machian) positivism is especially obvious in §160:

> The constitutional system shows that all objects can be constituted from "my elementary experiences" as basic elements; in other words (for this is what the expression "to constitute" means), all (scientific) statements can be transformed with retention of logical value into statements about my experiences (more precisely, about relations between them). *Every object* that is not

29 Again, the importance of this passage has been rightly emphasized by Sauer (compare footnote 21, this chapter): see (1985, §III; 1989, §V).

30 As noted above, there is no doubt that phenomenalist-empiricist motivations are explicitly expressed in the retrospective account of the *Aufbau* by Carnap (1963a). We shall return to these passages later, in §IV, this chapter.

> itself one of my experiences is thereby a quasi-object; its name is *an abbreviational auxiliary* [*abkürzendes Hilfsmittel*] for speaking about my experiences. Indeed, its name within constitutional theory and therefore within rational science is *only* an abbreviation....[31]

There is no doubt, then, that the constitutional system does realize the demand for a reduction or translation of all scientific statements in terms ultimately of the experiential given, and this feature of the constitutional system does clearly align it with the empiricist-positivist tradition.[32]

Carnap's official aim, however, is to represent neither the empiricist-positivist tradition nor the neo-Kantian tradition. For constitutional theory is officially neutral with respect to all disputes among different epistemological tendencies (§178):

> *the so-called epistemological tendencies of realism, idealism, and phenomenalism agree within the domain of epistemology. Constitutional theory presents the neutral basis [neutrale Fundament] common to all. They first diverge in the domain of metaphysics and thus (if they are to be epistemological tendencies) only as the result of a transgression of their boundaries.*

Since all epistemological tendencies agree that we begin with the data of experience and proceed from there via a "logical progress" to all other objects of cognition (see footnote 13, this chapter), disagreements arise only if we pose questions about which objects of cognition are "metaphysically real." But the constitutional system rejects "the metaphysical concept of reality" altogether (§176) and thus precisely represents what all epistemological tendencies agree upon while simultaneously rendering their remaining disagreements inexpressible. Insofar as there is a disagreement between the empiricist-positivist tradition and the neo-Kantian tradition, then, constitutional theory itself steadfastly refuses to take sides.[33]

31 For the relation between this conception and (Machian) positivism, see Carnap (1924, p. 109): "The positivistic philosophy, on the other hand, recognizes only the primary world; the secondary world is only an optional reorganization of the former, effected on grounds of economy."

32 Thus, although the neo-Kantian tradition begins with sensations and then seeks to embed them in a logico-mathematical "context," the idea of a logico-mathematical *reduction* to the given is entirely foreign to this tradition. Indeed, although he enthusiastically accepts the new mathematical logic as a formal tool, Cassirer (1910, Chapter II, §III; Chapter III, §III) just as emphatically rejects the logicist reduction of mathematics and prefers instead the axiomatic or formalistic conception associated with Hilbert. In this sense, Cassirer *rejects* Russell's "supreme maxim in scientific philosophizing."

33 See the continuation of the passage from §160 quoted earlier: "whether [the name of a quasi object] nevertheless still designates something 'subsisting in itself [*an sich Bestehendes*]' is a question of metaphysics, which can have no place within science (compare §161 and §176)."

Carnap's neutral and distant attitude toward the neo-Kantian tradition, in particular, is explicitly expressed very early on in the *Aufbau:*

> Are the constituted structures "generated in thought," as the Marburg School teaches, or "only recognized" by thought, as realism asserts? Constitutional theory employs a neutral language; according to it the structures are neither "generated" nor "recognized," but rather "*constituted*"; and it is already here to be expressly emphasized that this word "constitution" is always meant completely *neutrally*. From the point of view of constitutional theory the dispute involving "generation" versus "recognition" is therefore an idle linguistic dispute. (§5)

Yet it is not immediately clear what the force of Carnap's own distinction between "generation" and "constitution" is – especially since, as we have seen, Carnap says he agrees with transcendental idealism in §177 that all objects of cognition are "generated in thought."[34]

The title of §5 is "Concept and Object," and its main point is to argue that "the generality of a concept appears to us to be relative, and therefore the boundary between general concept and individual concept can be shifted in accordance with the point of view (see §158)." One important source of this relativity arises from what Carnap calls "the constitutional levels" (§40), namely, the type levels of *Principia Mathematica.* Thus, objects of any type other than the first appear as both classes (relative to objects of lower type) and objects or individuals (relative to objects of higher types). Carnap characterizes a class of objects as a "quasi-object" relative to its members and articulates "the *relativity of the concept 'quasi-object,'* which holds of an object of any constitutional level in relation to the objects of preceding levels." In the terminology of §5, therefore, "to every concept there belongs one and only one object, 'its object' (not to be confused with the objects falling *under*

See also the discussion by Carnap (1924). Carnap there outlines two different answers to the question of which is the "real" world. Realism and (transcendental) idealism hold that the secondary world is "reality," whereas positivism limits "reality" to the primary world (footnote 31, this chapter). Carnap (1924, pp. 109–10) then rejects the question: "We leave aside this properly speaking transcendent question of metaphysics; our immanent account has only to do with the character of experience itself, in particular with the distinction of its form-factors into necessary and optional, which we call primary and secondary, and with the relations between the two types. Here the *expression 'fiction'* also bears no metaphysical-negative value-character, but means that in our construction certain form-factors are newly added: the construction takes place 'as if' these factors belonged necessarily to experience, as primary." A similar rejection of "the metaphysical problem of reality" is found at the end of the manuscript "Vom Chaos zur Wirklichkeit" (footnote 10, this chapter).

34 For this reason, among others to be discussed below, Sauer (1985, §IV) finds it difficult to take seriously Carnap's attempt to distance himself from neo-Kantianism here.

the concept." The concept is the object viewed as a class, and thus in relation to lower levels in the hierarchy of types, whereas the corresponding object is the concept viewed as a member of classes in turn, and thus in relation to higher levels in the hierarchy of types. The structures thereby constituted can thus be viewed indifferently as either concepts or objects.

In §41, Carnap applies this relativity to philosophical-ontological distinctions among various "modes of being [*Seinsarten*]," specifically, to the distinction between "being and holding [*Sein und Gelten*]":

> Fundamentally, the *distinction between that which has being and that which has validity* [*dem Seienden und dem Geltenden*], which has been much emphasized in modern philosophy, also traces back to the distinction of object spheres – more precisely, to the distinction between proper objects and quasi-objects. Namely, if a quasi-object is constituted on the basis of certain elements of its domain that it "holds [*gilt*]" for these elements; thereby it is distinguished as having validity [*als Geltendes*] from the elements as having being [*als Seienden*].

Since, as §40 has shown, the distinction between proper object and quasi object is a relative one, it now follows that the distinction between being and holding is relative as well: "The concepts of being and holding are therefore relative and express the relation of each constitutional level to the immediately following one."

Now we saw in §II, this chapter, that a sharp distinction between being and holding, between the realm of concepts and the realm of objects, between the real or actual and the ideal or nonactual, is central to neo-Kantian epistemology. The real world of experience is contrasted with the ideal world of thought, whose "mode of being" consists in validity [*Geltung*] or necessity of thought [*Denknotwendigkeit*] rather than existence or actuality. Nevertheless, the real world of experience is made possible by the ideal world of thought; for only so are *objective judgments* about the real objects of experience possible. The idea is expressed succinctly in the following passage from Rickert:

> We therefore arrive at two worlds: a world of being [*einer seienden*] and a world of validity [*einer geltenden*]. But between them stands the theoretical subject, which combines the two through its judgements – whose essence is only understandable in this way – and without which we would not even be able to speak sensibly about existent [*seienden*] or real "objects" of cognition. (1892 [3rd ed. 1915], p. ix)[35]

35 In later editions, Rickert changes his terminology so that "being [*Sein*]" is now used comprehensively for both the real and the ideal realms. However, the "being and holding"

A quasi-Platonic, ontological distinction between real and ideal worlds – between the realm of being and the realm of validity – is thus fundamental to the neo-Kantian conception of the objects of cognition.[36]

In §41 of the *Aufbau*, by contrast, the distinction in question has been completely deflated. A philosophical distinction between two sharply separated "modes of being" has been transformed into a purely logical and explicitly relative distinction between an arbitrary rank in the hierarchy of types and the immediately succeeding rank. No traditional epistemological or ontological question is involved in the latter distinction, and this explains why Carnap, in §5, asserts that the closely related distinction between concept and object is a matter of complete indifference:

> Whether a certain object-sign means the concept or the object, whether a proposition holds for concepts or for objects, signifies no logical distinction but at most a psychological one – namely, a distinction in the representing ideas [*repräsentierenden Vorstellungen*]. In principle, there is absolutely no question of two different conceptions but only of two different interpretive manners of speaking. In constitutional theory we therefore sometimes speak of constituted objects, sometimes of constituted concepts, without making an essential distinction.
>
> These two parallel languages, which speak of objects and of concepts and still say the same thing, are fundamentally the *languages of realism and idealism.*

This is the precise sense in which constitutional theory is indeed neutral between realism and idealism, and hence between "recognition" and "generation in thought."[37] Accordingly, when Carnap explicitly articulates his

terminology is used even in the fifth (1929) edition of Rickert (1902). Cassirer (1910/1923, Chapter VII) also uses the "being and holding" terminology. Bauch (1923, Part I, Chapter II, §4) employs a more complicated terminology involving a further distinction between "holding [*Geltung*]" and "validity [*Gültigkeit*]." We observed earlier (footnote 23, this chapter) that the "being and holding" terminology goes back to Lotze.

36 The neo-Kantians characteristically take both Kant and Plato as models for their enterprise. For example, the Introduction to Rickert (1892) asks that we "venture onto the difficult and thorny path of logic and epistemology taken by Plato and Kant." See also Bauch (1923, Part I, Chapter IV, §5), Cassirer (1910/1923, Chapter IV, §III; Chapter VII, §I; Chapter VIII, §I). Compare also the discussion of Plato by Lotze (1874/1884, §§317–21). (Note added in 1998: It now seems to me that this attribution of a quasi-Platonic ontological distinction to Cassirer, in particular, is fundamentally misleading. Indeed, in some of the passages just cited, he endeavors explicitly to distance himself from such a view. For further discussion, see the Postscript to this chapter.)

37 By the same token, the distinction between real and ideal objects is also reconstructed within the constitutional system via purely formal-logical devices (§§158, 170–4): it rests in the end on formal-logical properties of the spatial and temporal orderings (themselves constituted as just two more types of orderings within the constitutional system) that make

points of agreement with "transcendental idealism" in §177, he is careful to maintain his neutrality: "all objects of cognition are constituted (*in idealistic language*, are 'generated in thought')" (emphasis added).

Carnap's distinction between realistic and idealistic languages is in fact the key to his philosophical neutrality. In §95, Carnap explains that the constitutional system can be presented in four different languages. However, "[*t*]*he fundamental language of the constitutional system is the symbolic language of logistics*" – that is, the language of *Principia Mathematica* (§107) – while "[t]he remaining three languages only provide translations of the logical fundamental language." The three remaining languages are then "a simple *translation in words*," "the translation in the *realistic language*," and finally "the *language of a fictional construction* [*Sprache einer fiktiven Konstruktion*].[38] In the last language, we view the strictly logical constitutional definitions (first language) "*as operational rules for a constructive procedure*," whereby "we have the task of prescribing for a given subject, designated as A, step by step operations through which A can arrive at certain schemata (the 'inventory-lists') corresponding to the individual objects to be constituted (§102)" (§99).

In the language of a fictional construction, we thus represent our subject A as undertaking a "synthesis of cognition [*Erkenntnissynthese*]" starting from the "given" (§100), on the basis of "synthetic components, and thus the constitutional forms" (§101). Since "[b]y *categories* are understood the forms of synthesis of the manifold of intuition to unity of the object," and since "[t]he manifold of intuition is called in constitutional theory 'the given,' 'the basic elements'" while "[t]he synthesis of this manifold to unity of an object is here designated as constitution of the object from the given" (§83), it follows that we can, if we like, view our subject A as undertaking a "synthesis of cognition" via "categories." There can be little doubt, then, that the language of a fictional construction is precisely the language of (transcendental) idealism. Whereas in the realistic language we view our constitutional definitions as capturing or representing independently given objects – the familiar objects of the empirical sciences (§98; compare §§52,

them particularly useful "as principles of individuation and therefore also of actualization [*Wirklichkeitssetzungen*] (which according to its meaning presupposes individuation)" (§158). In this way, the neo-Kantian ontological distinction between the real and the ideal is also logically deflated.

38 George consistently translates "*Konstitution*" as "construction," so he is here forced to translate "*fiktiven Konstruktion*" as "fictitious constructive operation." Since, as I will argue momentarily, the "language of a fictional construction" turns out to be just the language of "transcendental idealism," George's translation obscures precisely the distinction between "constitution" and "generation in thought" that Carnap is attempting to maintain. Compare footnote 1, this chapter.

75, 178), in the idealistic language we view our constitutional definitions as synthesizing or generating objects via the operations or constructions of a given cognitive subject.

The important point here, however, is that for Carnap the language of a fictional construction is indeed purely "auxiliary" or "fictional." The cognitive subject A, the step-by-step construction from the given, and the operations or acts of synthesis are all, strictly speaking, fictions, by which the underlying constitutional definitions are heuristically expressed "as palpable processes" (§99): "It is to be emphasized that *the constitutional system itself has nothing to do with these fictions*; they are referred only to the fourth language, and this serves only the didactic purpose of illustration." Similarly, although Carnap intends to give a rational reconstruction of the actual (empirical) process of cognition, he is careful to point out that the constitutional system itself involves no psychological processes whatsoever:

> Since the constitution indicates this function [a particular psycho-physical correlation] the course of the process of cognition is not somehow falsely presented through the constitution (namely, as a rational-discursive [process] instead of an intuitive one). (The latter occurs only in the language of a fictional construction, which can be given alongside as an intuitive aid.) The constitution itself indicates no process at all, but only the logical function in question. (§143)

For Carnap the fundamental language is always the purely formal-logical language of *Principia Mathematica*, wherein no cognitive subjects, no synthetic processes, and no acts or operations of construction are, in fact, to be found. On the contrary, in the strict "constitutional language" (§52), we have only a purely logical sequence of definitions formulated in a type-theoretic language containing a single nonlogical primitive.

For the neo-Kantians, by contrast, the language of cognitive subjects, synthetic processes, and acts or operations of thought is in no way dispensable. This is particularly evident in Cassirer's polemic against the antipsychologistic, mind-independent conception of logic and mathematics articulated in Russell's *Principles of Mathematics*. Cassirer (1910/1923, Chapter VII, §II) puts forward a "genetic" view of cognition on which (p. 315) "[i]t realizes itself only in a succession of logical acts, in a series, which must be successively run through in order that we become conscious of the rule of its progress." Accordingly, he explicitly opposes Russell's attempt wholly to remove the concept of the thinking subject from the realm of pure logic and mathematics so that (p. 316) "all closer relation of the ideal truths of mathematics and logic to the activity of thought falls away." Cassirer holds, on the contrary, that a "movement of thought" or an "act of production" is necessarily required (p. 317).

Indeed, from the point of view of the neo-Kantians, it is clear why this must be so. We start with a fundamental ontological distinction between the essentially timeless realm of validity or pure thought and the essentially temporal realm of reality or actuality; and we hold, moreover, that the latter is made possible in virtue of its "peculiar interweaving" with the former. We therefore need an intermediary standing between the two realms, as it were, and this is precisely the "thinking subject" (compare the above quotation from Rickert) – through which reality is "generated" by a succession of "logical acts." Thus, although the neo-Kantians explicitly oppose psychologism with respect to the realm of pure thought or pure logic [*reine Logik*], they nonetheless embrace an essentially psychological element in their account of the process of cognition. We are not involved here with *empirical* psychology, of course, but rather with the "transcendental psychology" of the "transcendental subject."[39]

Yet the *Aufbau*, as we have seen, dissolves the fundamental ontological distinction of the neo-Kantians into the purely formal-logical distinction between objects and quasi objects. Accordingly, Carnap himself has no need whatsoever for the "transcendental subject." Pure formal logic suffices to ground the objectivity of cognition all by itself, and no additional "transcendental psychology" is required:

> Many of the current objections to the autopsychological basis (or to "methodological solipsism") may be explained in terms of the failure to recognize this fact and this way [that science deals with logical structure and therefore objectivity is still possible in the context of an autopsychological basis] – and perhaps also many other formulations for the initial subject, such as, e.g., "transcendental subject," "cognitive subject," "trans-individual consciousness,"

39 See Rickert (1892, Chapters III–IV). Rickert speaks of the "theoretical subject," the "epistemological subject," the "judging subject," and "consciousness in general" rather than the "transcendental subject." Rickert (1892, Chapter IV, §§VI–VII) explicitly argues that "pure" logic is not sufficient to account for cognition; it must be supplemented precisely by "transcendental psychology." And this is also why Cassirer (1910/1923, Chapter VIII) feels compelled to add a final chapter on "the psychology of relations." Cassirer (1907, §IV) argues in the same vein that the merely "formal" logic of Russell and Couterat must be supplemented by "transcendental logic" in epistemology. Sauer (1985, §II; 1989, §IV) seems to me to give insufficient weight to the circumstance that Cassirer's conception of logic thus involves an essentially psychological element whereas Carnap's involves no such element. (Note added in 1998: Here again Cassirer's position is not sufficiently differentiated from Rickert's, this time with respect to "psychologism." See the Postscript to this chapter for further discussion.)

> "consciousness in general," which are perhaps to be interpreted as makeshift expedients [*Notbehelfe*], since one saw no way to the intersubjective [proceeding] from the natural initial point in the sense of an epistemological order of objects, namely from the autopsychological (compare the citations in §64). (§66)

And for Carnap, of course, the "natural initial point" is entirely "subjectless" (§65); an epistemological subject appears only in the language of a fictional construction. Once again, in the logically strict constitutional language, we have only a purely formal sequence of definitions – a sequence that happens to begin with a nonlogical primitive belonging to the domain of the autopsychological. Beyond this the constitutional system has no more intrinsic connection with psychology than it has with any other empirical science (compare footnote 12, this chapter).

In sum, the relationship between the *Aufbau* and neo-Kantian epistemology can perhaps best be expressed as follows. The neo-Kantians begin with an explicitly antipsychologistic conception of pure thought or pure logic intended to ground the objectivity of empirical cognition. The epistemological motivations of the *Aufbau* are in very substantial agreement with this idea. When the neo-Kantians put this idea into effect, however, the result is a fundamental philosophical-ontological distinction between two "modes of being" and a corresponding transcendental-psychological account of the "acts of synthesis" of the cognitive subject. For Carnap, by contrast, the notion of pure thought or pure logic is epitomized rather by the new mathematical logic of Frege and Russell, now considered as a powerful vehicle for dissolving philosophical confusions via logico-mathematical construction (compare footnote 32, this chapter). Carnap has been especially impressed, in particular, by Russell's conception, articulated in the second chapter of *Our Knowledge of the External World*, of "Logic as the Essence of Philosophy."[40] Epistemology in the *Aufbau* therefore becomes a logico-mathematical constructive project rather than a philosophical project in the traditional sense, and this logico-mathematical project is then a *replacement* for traditional epistemology (see the Preface to the first edition). Carnap's claim to complete philosophical neutrality in the end cuts deeply indeed.[41]

40 See Carnap's discussion of the impact of Russell's book on him in 1921 in Carnap (1963a, p. 13). Russell's new conception of "the logical-analytic method of philosophy" is clearly the central point.

41 Sauer (1985, §IV) explicitly rejects Carnap's claim to philosophical neutrality, arguing that Carnap is in fact committed either to neo-Kantian epistemology *simpliciter* or to (Machian) positivism. In particular, the only way in which Carnap can distance himself from

IV

The preceding analysis results in a very different picture of the epistemological significance of the *Aufbau* than that which has been customary within contemporary Anglo-American philosophy. The *Aufbau* is not best understood as starting from fundamentally empiricist philosophical motivations and then attempting to put these into effect – on the basis of the new mathematical logic of *Principia Mathematica* – in a more precise and rigorous way than had been previously possible. The epistemological motivations of the *Aufbau* begin rather with the concerns and problems of the neo-Kantian tradition: with a concern for depicting how the cognitive process transforms inherently private and subjective sensations into fully objective experience capable of validity and truth, and with the corresponding problem of carrying out this project in an essentially "logical" – that is, nonmetaphysical and nonpsychological – fashion. Yet the neo-Kantians themselves (for present purposes, Rickert, Bauch, and Cassirer) are not able fully to achieve such a purely logical standpoint, principally because they are not in possession of a sufficiently rich and determinate conception of logic itself. The neo-Kantians are instead left in the somewhat uncomfortable position – almost in spite of themselves, as it were – of appealing to ontological distinctions among various "modes of being" and psychological accounts of the activities of the cognitive subject. In the *Aufbau*, however, the new mathematical logic of *Principia Mathematica* provides Carnap with all the philosophical concepts and distinctions he needs. Carnap thereby achieves a standpoint that is both nonpsychological and truly metaphysically neutral, and, at the same time, he transforms the neo-Kantian tradition into something essentially new: "logical-analytic" philosophy.[42]

neo-Kantian epistemology, according to Sauer, is by embracing a positivist conception on which "only the elementary experiences possess the character of reality" – a conception Sauer finds expressed in Carnap's claim that all objects except the elementary experiences are quasi objects (§160). I have tried to show, on the contrary, how Carnap is otherwise distanced from neo-Kantian epistemology (compare footnotes 34 and 39). Sauer's reading of §160 and Carnap's relation to positivist epistemology seems to me to be equally mistaken. For Carnap's notion of "reality" is also constructed *within* the constitutional system: "In the same sense *the expression 'quasi-object' designates only a determinate logical relationship*, not the negation of a metaphysical reality-value. In fact all real objects (they are recognized in constitutional theory in the same sphere as real as in the factual sciences, see §170) are quasi-objects" (§52; compare footnotes 33 and 37, this chapter).

42 In this conception of the relationship between the *Aufbau* and neo-Kantianism, I have been especially helped by discussions with Alan Richardson – which is not to say that he would fully agree with it of course. For his perspective on this issue, see Richardson (1992).

This picture of the epistemological significance of the *Aufbau* harmonizes particularly well with Carnap's early philosophical development.[43] Carnap's first published work, *Der Raum* (1922) (his doctoral dissertation), explicitly articulates a modified Kantian conception of space.[44] Kant was wrong, to be sure, in thinking that three-dimensional Euclidean space is an a priori necessary condition of the possibility of experience. Nevertheless, Kant was perfectly correct about the experience-constituting function of space – it is just that a more general structure is required:

> It has already been explained more than once, from both mathematical and philosophical points of view, that Kant's contention concerning the significance of space for experience is not shaken by the theory of non-Euclidean spaces, but must be transferred from the three dimensional Euclidean structure, which was alone known to him, to a more general structure.... According to the foregoing reflections, the Kantian conception must be accepted. And, indeed, the spatial structure possessing experience-constituting significance (in place of that supposed by Kant) can be precisely specified as topological intuitive space with indefinitely many dimensions. We thereby declare, not only the determinations of this structure, but at the same time those of its form of order [*n*-dimensional topological *formal* space] to be conditions of the possibility of any object of experience whatsoever. (1922, p. 67)

Accordingly, the main point of *Der Raum* is to show how the contemporary philosophical disputes about the nature of space and geometry can be dissolved by distinguishing among three distinct "meanings" of space. Formal space is a purely logical structure constructed within the theory of relations or theory of order; it therefore has the formal or analytic character defended by such thinkers as Russell and Couterat (1922, Chapter I). Intuitive space, by contrast, is given to us by a kind of nonformal (but also noninductive) sensory procedure; here we, in fact, have the Kantian synthetic a priori, but only *infinitesimally* Euclidean properties (topological properties sufficient to admit some or another Riemannian metrical structure) are thereby intuitively given (1922, Chapter II). Physical space, finally, is constructed by fitting actual empirical data into the already given space of intuition; here we conventionally choose a particular determinate metrical structure for space

43 Here I am again in very substantial agreement with Sauer (1989, §III). Richardson (1998) provides an especially rich and detailed account of Carnap's early period; see also Richardson (1992).

44 See Chapter 2 (this volume) for more detailed discussion.

and also (in accordance with the empirical data and the methodological principle of simplicity) a particular determinate dimension number (1922, Chapter III).

Carnap's early conception is therefore characterized by two different levels of nonempirical, experience-constituting structure. Topological structure is necessary and unique, but metrical structure is subject to conventional choice from among a wide spectrum of alternatives (all the geometries definable within Riemann's theory of manifolds). Carnap (1922, pp. 38–40) marks this distinction by a "division within the realm of form between necessary and optional [*wahlfreier*] form": the former comprises the topological structure without which no experience at all is possible; the latter comprises the particular metrical geometry (and dimension number) freely chosen on conventional (and methodological) grounds. The point is that some or another metrical structure is indeed necessary for fully objective (scientific) experience, but no particular such structure is a priori given. The metrical structure of physical space – precisely because it, too, is constitutive of experience – cannot be determined empirically but is instead entirely up to our free choice.[45]

The distinction between necessary and optional form is then applied by Carnap (1924), where the first hint of an *Aufbau*-style construction of the external world from the "given" appears in print (see footnote 10, this chapter). Carnap's aim there is to explain how the "secondary world" of physical objects arises from the "primary world" of immediate sensations on the basis of the "fictions" of three-dimensional space and thoroughgoing causal determination; and Carnap begins by situating his project in relation to neo-Kantianism in a particularly striking fashion:

> The neo-Kantian philosophy is not acquainted with the primary world, since their conception that the forms of experience of [the secondary world] are necessary and unique prevents them from recognizing the distinction between the primary and the secondary world. Their true achievement, namely, the demonstration of the object-generating function of thought, remains

45 For the antiempiricist character of Carnap's conventionalism of this period, compare again the passage from Carnap (1923) quoted in footnote 19, this chapter. Sympathy with the Kantian conception of space (and time) is expressed as late as Carnap (1925, p. 231): "The 'external world' surrounding us displays two types of order: that of succession [*Nacheinander*] and that of coexistence [*Nebeneinander*]. Since Kant we customarily answer the question of why every object of (outer) experience fits into these orders by the idea that they are forms of intuition and therefore conditions to which every object must conform in order in general to be object of a possible experience."

untouched, however, and underlies our conception of the secondary world as well. (1924, p. 108)

The first sentence of this passage may well suggest that Carnap is aligning himself with positivist-empiricist epistemology here.[46] It quickly becomes clear, however, that Carnap's conception of the "primary world" does not rest on an empiricist preoccupation with the certainty and foundational role of immediate sensory experience, but rather on precisely the distinction between necessary and optional form previously articulated in *Der Raum*.[47] Carnap's complaint against the neo-Kantians is simply that they fail to recognize the importance of conventional – and thus freely chosen or optional – factors within "the object-generating function of thought."[48]

There can be little doubt, therefore, that the project of the *Aufbau*, although by no means entirely independent of the empiricist-positivist tradition, originates within a primarily Kantian and neo-Kantian philosophical context. Experience is to be constituted from sensation on the basis of forms imposed by thought, but these forms are increasingly deprived of the fixed, synthetic a priori character ascribed to them by Kant. Indeed, in the *Aufbau* itself, no remaining trace of the synthetic a priori can be found.[49] In particular, Carnap makes no distinction between necessary (intuitive) and optional (conventional) form: all form is now purely *logical*, and only conventions and the analytic a priori remain.[50] Accordingly, in the *Aufbau* the "primary

46 See footnote 31, this chapter, and Carnap (1924, p. 106): "The critique that has been exerted on the Kantian concept of experience, especially from the positivist side, has taught us that it is by no means the case that all form-factors in experience to which *Kant* ascribes necessity actually possess it."

47 See Carnap (1924, pp. 106–7, 109–10), and compare footnote 33, this chapter.

48 This point, in particular, is well emphasized by Sauer (1989, p. 115). Sauer also suggests that Carnap's disagreement with neo-Kantianism here is more apparent than real, since the Marburg School (Cassirer especially) was in fact willing to liberalize (i.e., *relativize*) the synthetic a priori and to admit alternative (e.g., non-Euclidean) experience-constituting structures. Sauer is, of course, perfectly correct with respect to the Marburg School, but the Southwest School of Rickert and Bauch was considerably more rigid. Rickert (1892, Chapter V) defends the idea that only the spatiotemporal-causal (and presumably Euclidean) world of common sense can constitute "reality," whereas Bauch (1923, pp. 260–2) had to be convinced by Cassirer (1921) that non-Euclidean as well as Euclidean geometries could play an experience-constituting role. It is clear from the passage cited in footnote 46, this chapter, that Carnap has the more rigid (and thus more genuinely Kantian) versions of neo-Kantianism in mind here.

49 Therefore, I cannot follow Sauer (1989, p. 114) in the idea that "[Carnap's] doctrine of synthetic *a priori* forms of experience, however, he abandoned only after he had come to Vienna in 1926."

50 Compare the discussion of conventions in the *Aufbau* in §I, this chapter – particularly, footnotes 17, 19, and 20. The problem of relating conventions, on the one hand, and the logic

world" has even less of a position of epistemic privilege than it had in Carnap (1924). The autopsychological realm and the physical realm are no longer marked off from one another by the distinction between necessary and optional form but simply in virtue of their purely logical differences as distinct "object-spheres" within a type-theoretic hierarchy (compare §132 and footnote 18, this chapter). And this last is, of course, just the kind of revolutionary philosophical move to which we have called attention earlier – a move by which Carnap simultaneously distances himself from both the neo-Kantian tradition and the empiricist-positivist tradition.

Yet, as we saw in §I (this chapter), Carnap (1963a) retrospectively describes the motivations of the *Aufbau* in the most explicitly empiricist and phenomenalist terms imaginable. There is "a certain rock bottom of knowledge, the knowledge of the immediately given," and "the task of philosophy consists in reducing all knowledge to a basis of certainty." Therefore "a phenomenalistic language [is] the best for a philosophical analysis of knowledge." Since, as we also saw in §I, such a phenomenalist-foundationalist conception is hardly in evidence in the text of the *Aufbau* itself, and, as we have just seen, this is equally true of Carnap's pre-*Aufbau* writings, Carnap's retrospective account is puzzling indeed. It is not unprecedented, of course, for the character and motivations of an earlier and now rejected philosophical project to be grossly misdescribed – even by the philosopher whose earlier views are in question. Nevertheless, some explanation is still required, at the very least, for why Carnap has chosen here to describe his earlier views in precisely these terms.

My suggestion is that, in the passages in question from his "Intellectual Autobiography," Carnap is describing not so much his own motivations when writing the *Aufbau* but rather the way in which the *Aufbau* was initially understood within the Vienna Circle. That this may well be the case is already indicated by the contexts in which each of the two passages occur; for both occur in discussions of how the (majority of the) Circle moved away from a preference for a phenomenalistic language and, under Neurath's influence, toward a preference for a physicalistic language. Thus the first passage occurs in a section on "Physicalism and the Unity of Science" and is bracketed by the following sentences:

> In our [the Vienna Circle's] discussions we were especially interested in the question whether a phenomenalistic language or a physicalistic language was

of *Principia Mathematica*, on the other, remains unsolved here, however. Although Carnap gestures toward a conventionalist conception of logic in §107 of the *Aufbau*, he is not able coherently to articulate such a conception until *Logische Syntax der Sprache* in 1934.

> preferable for the purposes of philosophy.... In the Vienna discussions my attitude changed gradually toward a preference for the physicalistic language. (1963a, p. 50)

A discussion of Neurath's arguments for physicalism occupies the next two pages. The second passage occurs in a section on "Liberalization of Empiricism" and is introduced as follows:

> The simplicity and coherence of the system of knowledge, as most of us in the Vienna Circle conceived it, gave it a certain appeal and strength in the face of criticisms. On the other hand, these features caused a certain rigidity, so that we were compelled to make some radical changes in order to do justice to the open character and inevitable uncertainty of all factual knowledge.
>
> According to the original conception, the system of knowledge, although growing constantly more comprehensive, was regarded as a closed system in the following sense. We assumed that there was a certain rock bottom of knowledge.... (1963a, pp. 56–7)

Neurath's (and Popper's) arguments against such a "rock bottom of knowledge" are then discussed in some detail. Here Carnap is, of course, describing the well-known protocol-sentence debate, which split the Vienna Circle into a "left wing" (antifoundationalism), represented by Neurath, Hahn, and Carnap, and a "right wing" (foundationalism), represented by Waismann and Schlick.[51]

Now, with respect to the *Aufbau* itself, we know that Carnap completed a first draft in 1922–5 while living in Jena and Buchenbach. He became acquainted with Schlick in the summer of 1924 (through Reichenbach) and lectured on the *Aufbau* in Vienna to Schlick's Philosophical Circle in 1925. In 1926, Carnap joined the University of Vienna, and a typescript of the first version of the *Aufbau* was read and intensively discussed at meetings of the Circle.[52] Carnap describes the initial reception as follows:

> From the very beginning, when in 1925 I explained in the Circle the general plan and method of *Der logische Aufbau*, I found a lively interest. When I returned to Vienna in 1926, the typescript of the first version of the book was read by the members of the Circle, and many of its problems were thoroughly discussed. Especially the mathematician Hans Hahn, who was strongly

51 For detailed discussion of this debate, see Uebel (1992) and Oberdan (1993).

52 See Carnap (1963a, pp. 10–20). A revised version of the *Aufbau* then was published in 1928. It would, of course, be fascinating and most relevant here to compare the 1925–6 typescript with the published version. Unfortunately, it has so far proved impossible to locate a copy of the typescript – either in the Pittsburgh Archives for Scientific Philosophy, among Carnap's papers at the University of California at Los Angeles, or at the University of Vienna.

> interested in symbolic logic, said that he had always hoped that somebody would carry out Russell's program of an exact philosophical method using the means of symbolic logic, and welcomed my book as the fulfillment of these hopes. Hahn was strongly influenced by Ernst Mach's phenomenalism, and therefore recognized the importance of the reduction of scientific concepts to a phenomenalistic basis, which I had attempted in the book. (1963, p. 20)

It appears, then, that at Vienna the *Aufbau* was introduced into a philosophical context already predisposed toward phenomenalistic empiricism.

What we know about the background and early history of the Vienna Circle is entirely consistent with this idea. Ernst Mach himself was the first occupant of the chair for philosophy of the inductive sciences at the University of Vienna from 1895 to 1901. In the years 1907–12, a group of Viennese scientific thinkers deeply influenced by Mach (as well as by the French "neopositivists" Duhem, Poincaré, and Rey) met regularly to discuss philosophy of science and "the decline of mechanism." This group was led by Philipp Frank, Otto Neurath, and Hans Hahn, and has been dubbed the "First Vienna Circle."[53] Although the First Vienna Circle ceased to meet regularly in 1912, when Frank went to Prague, its members remained in contact and continued to pursue and defend Machian ideas. Indeed, they saw Einstein's construction in 1915–16 of a relativistic theory of gravitation as a particularly striking proof of the scientific fruitfulness of Mach's positivist critique.[54] Schlick, at Rostock and Kiel, was at the same time moving toward a very similar understanding of the philosophical significance of Einstein's theory.[55] The result was that Hahn succeeded in bringing Schlick to Vienna to occupy the chair for the philosophy of the inductive sciences in 1922, and what we now know as the Vienna Circle was born.[56]

When Carnap was brought to Vienna by Schlick in 1926, he therefore found himself in a philosophical climate within which the phenomenalist-empiricist aspects of the *Aufbau* were bound to be given the most prominent emphasis. This still does not account for the foundationalist concern with certainty and a "rock bottom of knowledge" manifest in our passages from

53 See Haller (1985/1991). See also the Introduction to Frank (1949), and Kraft (1950/1953). Kraft was, of course, an original member of Schlick's Philosophical Circle.

54 See, e.g., Frank (1949, p. 68).

55 Schlick (1918/1985) – clearly under the influence of his teacher Planck – sharply attacks Machian phenomenalism (together with Russell's external world program). However, Schlick's evolving assimilation of Einstein's general relativity theory led him progressively closer to Mach – particularly by 1921–2. See Chapter 1 (this volume) for further discussion.

56 See Frank (1949, pp. 32–3). Of course the Vienna Circle was not established as an official society, under the rubric of the *Verein Ernst Mach*, until 1928.

Carnap (1963a), however, because Machian phenomenalism does not itself amount to foundationalism.[57] The most important factor responsible for such a foundationalist reading of Machian phenomenalism – and therefore of the *Aufbau* as well – was undoubtedly the assimilation within the Circle of Wittgenstein's *Tractatus.* For the Circle understood the *Tractatus* as articulating a foundationalist-empiricist conception of *meaning.* Definitions explain the meanings of words in terms of other words, but this procedure cannot go on to infinity, or else no word ultimately has meaning at all. Therefore, all meaning must finally rest on primitive acts of ostension, and what is ostended must be immediately given:

> Definitions are ultimately reducible to ostension of what is designated. One can point only at something which is immediately given, and thus only at what is perceivable. In this way, what assertions can possibly mean is tied to experience. No meaning can be given to that which is not reducible to experience; and this is a consequence of fundamental importance. (Kraft 1950/1953, pp. 32–3)[58]

And there is no doubt that this conception of meaning – and this understanding of the *Tractatus* – was adopted especially by Waismann and Schlick.[59] It was then entirely natural to read Carnap's *Aufbau* as the precise realization of just such a "Tractarian" theory of meaning.[60]

57 Very little evidence of such epistemological concerns is found in Mach, who instead tended to disavow all purely philosophical motivations and rest his case rather on the unity of science – the need, in particular, to unify physics and psychology. Moreover, this is certainly how Philipp Frank (and presumably other members of the First Vienna Circle) read Mach: compare Frank (1949, Chapter 3).

58 Kraft (pp. 30–1) attributes the origin of this conception – which of course leads naturally to the verifiability theory of meaning – to Wittgenstein. Interestingly enough, Kraft uses the very same conception of ultimately ostensive meaning to introduce his exposition of Carnap's constitutional system (pp. 83–4), whereas, as we know from §II, this chapter, Carnap himself here explicitly rejects ostension as a source of meaning (see §§13, 16). Kraft is not unaware of this circumstance, and he handles the conflict by distinguishing between *subjective* meaning based on ostension and *intersubjective* meaning based on logical structure (p. 84). But the important point is that Kraft's notion of *subjective* meaning is entirely foreign to the *Aufbau.*

59 Waismann's *Theses* (circa 1930), which purport to summarize the *Tractatus,* explicitly articulate this conception in §7, on "Definition" – see (Waismann (1967/1979, pp. 246–53). Schlick puts forward essentially the same argument in virtually all of his papers from the early 1930s; for discussion see Chapter 1) (this volume).

60 Again, see Kraft (1950/1953, pp. 114–17, especially p. 117): "Wittgenstein identified [atomic propositions] with the propositions he called 'elementary propositions.' They are propositions which can be *immediately* compared with reality, i.e., with the data of experience. Such propositions must exist, for otherwise language would be unrelated to reality. All propositions which are not themselves elementary propositions are necessarily truth-functions of elementary propositions. Hence all empirical propositions must be reducible

The connection between this foundationalist conception of meaning and foundationalist epistemology is made explicitly by Schlick (1934/1979). In the opening paragraph, Schlick (p. 370) describes the traditional need of philosophy "to seek an unshakable foundation which is removed from all doubt and forms the firm ground on which the unsteady structure of our knowledge is erected," to seek "the natural bedrock [*natürlichen Felsen*] which exists *before* all building and does not itself totter." Schlick then finds such "bedrock" in his notorious "affirmations [*Konstatierungen*]," which ostensively and demonstratively report on a sensory content present here and now. But why exactly are such "affirmations" absolutely certain? As in the parallel case of analytic propositions, a knowledge of their truth is simply inseparable from a grasp of their meaning:

> In other words, I can understand the sense of an "affirmation" only when, and only whereby, I compare it with the facts and thus carry out that process that is required for the verification of all synthetic propositions. Whereas, however, in the case of all other synthetic assertions the establishing of the sense and the establishing of the truth are separate and easily distinguishable processes, in the case of observational propositions the two coincide – just as in the case of analytic propositions. Thus as different as the "affirmations" are from analytic propositions otherwise, they still have in common with them that in both cases the process of understanding is at the same time the process of verification – I grasp the truth at the same time as the sense. In the case of an affirmation it would have just as little sense to ask whether I could perhaps be mistaken about its truth as it would in the case of a tautology. Both are absolutely valid. Only the analytic proposition or tautology is at the same time empty of content, while the observational proposition provides us the satisfaction of genuine cognition of reality. (1934/1979, p. 385)

To grasp the meaning of an affirmation is to be ostensively confronted with the very fact whose existence it reports. Error is therefore impossible, and we have thus found the true foundation of all (synthetic) knowledge.

Carnap's own writings from this period exhibit a strikingly different character.[61] Although he, of course, also adopts the verifiability theory of meaning and acknowledges the importance of Wittgenstein's *Tractatus* in

to propositions about the given.... The reduction is made possible by a family-tree of concepts which exhibits their reducibility to relations between experiences, the sort of reduction sketched in Carnap's constitution-system. In this way the empiricist theories of meaning, concepts and propositions are all interconnected." Although there is a suggestion of the verifiability theory of meaning in §179 of the *Aufbau*, I am tempted to conjecture that this section was written after Carnap came to Vienna. Unfortunately, we are not in a position to verify this conjecture (see footnote 51, this chapter).

61 The writings that I have in mind here are Carnap (1928b, 1930a, 1932a,b,c,d).

this regard, Carnap does not embrace the empiricist foundationalism articulated by Waismann and Schlick. For, in the first place, the notion of ostension plays no role at all in his conception of meaning, which continues to be explained purely formally: to know the meaning of a sentence S is to know which sentences S is deducible from and which sentences are deducible from S.[62] To be sure, there is a special class of sentences, the protocol sentences, against which all other (nonanalytic) sentences are tested – by deducing sentences of this special class from sentences in the latter class. Nevertheless, in the second place, these protocol sentences need not be related to "immediate experience" in any antecedently understood sense. Indeed, Carnap explicitly leaves to one side the question of the content of such protocol sentences and, in particular, whether protocol sentences have the form of (a) Machian sensation reports, (b) *Aufbau* style reports of holistic elementary experiences, or (c) reports about ordinary observable things.[63] And there clearly can be no question of absolute certainty *á la* Schlick in case (c). It follows, in the third place, that the sense in which protocol sentences are epistemically privileged is also purely formal: other sentences are tested (and accordingly accepted or rejected) by the logical deduction of protocol sentences, but protocol sentences (trivially) are not so tested in turn. Protocol sentences are simply the logical termini of the procedure of testing or verification.[64]

The fourth point of difference between Carnap and the Schlick-Waismann conception is perhaps most significant of all. For Carnap continues to hold, as he did in the *Aufbau*, that there are two essentially distinct but equally useful ways of reconstructing the language of science. The first way, corresponding to the standpoint of "methodological positivism," consists in beginning with protocol sentences – considered as distinguishable from the rest of the

62 See Carnap (1932a, §2), and compare Carnap (1928b, Part I, §A.1). On ostension, see Carnap (1932b, §2), which asserts – contrary to the usual conception – that ostensive definitions involve translation *within* language just as much as do nominal definitions.

63 See Carnap (1932a, §2; 1932b, §3).

64 See Carnap (1932b, §3): "The simplest sentences of the *protocol language* are the protocol-sentences, i.e., sentences which do not themselves require confirmation [*Bewährung*] but serve as basis [*Grundlage*] for all other sentences of science." It is clear from the context, I think, that the sense in which protocol sentences do not *require* confirmation is simply that their truth or falsity is not in fact at issue in the procedures of scientific testing that Carnap aims here rationally to reconstruct. See also Carnap (1932c, §7): "The difference [between a "system sentence" – i.e., a scientifically testable sentence – and a protocol sentence] rests on the fact that the system sentence... may, under certain circumstances, be disavowed, whereas a protocol-sentence, being an epistemological point of departure, cannot be rejected." Again, the sense in which a protocol sentence *cannot* be rejected is simply that such sentences are not, in fact, at issue in the relevant process of testing.

language of science – and exhibiting the logical relations in virtue of which all other sentences of science are epistemically based on the protocol sentences. The second way, corresponding to the standpoint of "methodological materialism," consists in beginning with the basic language of physics and translating all other sentences – including the protocol sentences (which, from this point of view, appear simply as sentences of empirical psychology) – into the language of physics. Moreover, Carnap continues to hold that the materialistic or physicalistic system is the most appropriate system for representing the content of science as a completely unified and fully intersubjective body of knowledge:

> We speak of "methodological" positivism or materialism, respectively, because we are here concerned with only the method of conceptual derivation, while the metaphysical-positivistic thesis of the reality of the given and the metaphysical-materialistic thesis of the reality of the physical remain completely excluded here. Therefore, positivistic and materialistic constitutional systems do not contradict one another. Both are legitimate and unavoidable. The positivistic system corresponds to the epistemological viewpoint, because in it the validity of a cognition is shown through reduction to the given [*da sich in ihm die Gültigkeit einer Erkenntnis durch Rückführung auf das Gegebene erweist*]. The materialistic system corresponds to the standpoint of factual science [*Realwissenschaft*], because in it all concepts are reduced to the physical – to the only domain that exhibits thoroughgoing law-governedness and makes intersubjective knowledge possible. (1930a, §8)[65]

In the context of the protocol-sentence debate, Carnap therefore needed only the smallest push from Neurath to break decisively from the Schlick-Waismann wing – that is, from the "absolutism of the 'given,'" and even from the "refined absolutism of the primitive sentence" (1932d, §3).[66]

All the evidence then suggests that, when Carnap uses empiricist-foundationalist language in his "Intellectual Autobiography," he is not really

65 See also Carnap (1932b, §7), and compare §§133, 136 of the *Aufbau*. With respect to the issue raised in footnote 64 (this chapter), it should be pointed out that Levi's translation of our passage from Carnap (1930a) misleadingly suggests that Carnap has a more traditionally foundationalist conception of epistemic justification in mind. For Levi renders the fourth sentence thus (1930a/1959, p. 144): "The positivist conception corresponds to the epistemological viewpoint because it proves the validity of knowledge by reduction to the given." As I read Carnap here, he is saying that, just as in the constitutional system of the *Aufbau*, the point of the positivistic system is simply to display – rationally to reconstruct – the epistemological justification or validity that a cognition *already has* in virtue of its place in the system of the sciences.

66 The point is that once protocol sentences are translated into the physical language they no longer have even the "refined" epistemic privilege of footnote 64, this chapter.

describing his own views – and he is certainly not describing the actual motivations of the *Aufbau.* He is rather depicting the philosophical dialectic between the "right wing" and the "left wing" of the Circle in a particularly clear and dramatic fashion. Although there is no doubt that the *Aufbau* – particularly as read by Schlick and Waismann – played a central role in this dialectic, Carnap himself was never moved by epistemological foundationalism. Carnap's position did indeed change significantly after the confrontation with Neurath in the protocol-sentence debate, but this change did not consist in the abandonment of foundationalist epistemology in particular (to which he was never attracted in any case). What Carnap gave up was any interest in traditional epistemology and its rational reconstruction at all.[67] He instead came to see that the metaphysical neutrality that he sought throughout his philosophical career could best be achieved in the context of logical investigation into the formal structure of any and all constitutional systems – or, to use his later terminology, into the structure of any and all formal languages or linguistic frameworks. And this leads, once again, to the program of *Logische Syntax der Sprache.*

POSTSCRIPT: CARNAP AND THE NEO-KANTIANS

Sections II and III of this chapter present a misleading picture of the relationship between the *Aufbau* and neo-Kantian epistemology, largely because they too closely assimilate the views of the Southwest and Marburg Schools. The central point of difference between Carnap and neo-Kantianism generally is located in a quasi-Platonic conception, supposedly common to both schools, according to which the real or actual is dualistically opposed to the ideal or nonactual: "being [*Sein*]" is *ontologically* opposed to "validity [*Geltung*]." Carnap's attempt to distance himself from the Marburg School in §5 of the *Aufbau* is then read primarily in terms of §41, where Carnap explicitly deflates the ontological distinction between "being" and "validity" by reinterpreting it within the type hierarchy of *Principia Mathematica* as the purely logical distinction between objects and quasi objects. The problem with this reading (as indicated in footnote 36, this chapter) is that, whereas the Southwest School is indeed entangled with a quasi-Platonic ontological distinction between real and ideal realms – the realm of "being" and the realm of "validity" – the Marburg School explicitly rejects this kind of

67 This point is brought out especially well by Richardson (1998, pp. 207–13), via a discussion of the important transitional paper of Carnap (1936). Compare also Richardson (1996).

dualism. Indeed, according to the Marburg School, all such philosophical dualities have only a *relativized* meaning within an essentially unitary progress of knowledge – in a way, moreover, that is closely analogous to Carnap's own conception of the sequence of levels constituting the fundamentally unified cognitive process the *Aufbau* aims rationally to reconstruct.[68]

The most basic idea of the Marburg School is a "genetic" view of empirical knowledge. Empirical knowledge, in contradistinction to pure mathematical knowledge, consists of a methodological series or sequence of historically given mathematical structures or theories – a sequence of theories in mathematical physics, for example – which successively refine our attempts conceptually to master the phenomena. And what essentially distinguishes such a sequence of structures from those generated in pure mathematics is simply that the *empirical* methodological sequence is necessarily potentially infinite or nonterminating:

> In contrast to the mathematical concept, however, [in empirical science] the characteristic difference emerges that the construction [*Aufbau*], which within mathematics arrives at a fixed end, remains in principle *incompletable* within experience. No matter how many "layers [*Schichten*]" of relation we may erect one upon another, and however close we may come thereby to all particular circumstances of the real process, the possibility is still always open that some codetermining factor of the total result was left out of account and will only be discovered by the further progress of experimental analysis. Every conclusion that we here achieve has thus only the relative value of a preliminary fixation, which holds fast what is gained only in order to use it, at the same time, as a starting point for new determinations. (Cassirer 1910/1923, p. 254)

But this potentially infinite sequence of "determinations" or structures given in the progress of empirical knowledge must be conceived, at the same time, as necessarily converging:

> [Knowledge] realizes itself only in a succession of logical acts, in a series, which must be run through successively, so that we may become conscious of the rule of its progress. But if this series is to be grasped as a *unity*, as an expression of an *identical* state of affairs, which is thereby designated more sharply and precisely the further we advance, then it is required for this that we must think of the series as converging towards an ideal *limit*. The limit "is" and subsists [*besteht*] in univocal determinateness [*eindeutiger Bestimmtheit*], although for us

68 I here draw on Friedman (1996). I am indebted to Alan Richardson and Werner Sauer for rightly protesting against the assimilation of the Marburg and Southwest Schools in my earlier discussion in 1992.

> it is *attainable* in no other way than through the individual terms of the series and their lawful change. (p. 315)

Empirical reality itself, on this view, is thus the ideal limit point – the never completed "X" – toward which the methodological progress of science is converging. It is this ideal limit point, in the end, that constitutes the "object" of empirical knowledge.

Moreover, the fundamental error of all "metaphysical" conceptions of knowledge, on this view, is an hypostatization of successive stages in the methodological progress of empirical cognition via an *ontological* dualism of opposing elements:

> [T]he sense of all objective judgements reduces to a final original relation, which can be expressed in different formulations as the relation of "form" to "content," as the relation of "universal" to "particular," as the relation of "validity" to "being." Whatever designation one may finally choose here, what is alone decisive is that the basic relation itself is to be retained as a strictly *unitary* relation, which can only be designated through the two opposed moments that enter into it – but never constructed out of them, as if they were independent constituents present in themselves. The original relation is not to be defined in such a way that the "universal" somehow "subsisted [*bestände*]" next to or above the "particular" – the form somehow separate from the content – so that the two are then melded with one another by means of some or another fundamental synthesis of knowledge. Rather, the unity of mutual *determination* constitutes the absolutely first datum, behind which one can go back no further, and which can only be analyzed via the duality of two "viewpoints" in an artificially isolating process of abstraction. It is the basic flaw of all metaphysical epistemologies that they always attempt to reinterpret this duality of "moments" as a duality of "elements." (Cassirer 1913, pp. 13–14)[69]

This means, in particular, that all such fundamental dualities have only a *relativized* significance, which must always be understood in the context of some or another given stage of the never-ending methodological progress:

> Matter *is* only in relation to form, just as form *is valid* [*gilt*] only in relation to matter. If one neglects this coordination [*Zuordnung*] then there remains no "existence [*Dasein*]" for either – into whose ground and origin one could

69 This essay is a critical review of contemporary trends in epistemology, with particular attention to the views of Rickert and the Southwest School. I am indebted to Werner Sauer for first calling it to my attention and for emphasizing to me, in this connection especially, the crucially important differences between Cassirer and the Southwest School.

> inquire. The material particularity of the empirical content can therefore never be cited as a proof for the dependence of all cognition of objects on an absolutely "transcendent" ground of determination: for this determinateness, which as such undeniably subsists [*besteht*], is nothing other than a characteristic of *cognition itself*, through which its concept is first achieved. (Cassirer 1910/1923, p. 311)

In this sense, Cassirer's quite explicit attempt to deflate the fundamental ontological distinctions of the Southwest School closely parallels the Carnapian strategy of §41 of the *Aufbau*.[70]

By the same token, it is also a mistake to assimilate Cassirer's discussion of "the psychology of relations" to Rickert's appeal to a "transcendental psychology" (footnote 39, this chapter). Just as fundamental ontological dualities such as those between "form" and "matter," "universal" and "particular," and so on have meaning only *within* the unitary methodological progress of scientific knowledge, the same is true of the notion of "consciousness" itself:

> [W]ithout a *temporal* sequence and order of contents, without the possibility of collecting them into determinate *unities* and of separating them again into different *pluralities*, without the possibility, finally, of distinguishing relatively constant subsistent entities [*Bestände*] from relatively changing ones, the thought of the I has no specifiable meaning and application. Analysis teaches us with univocal determinateness *that* all these relational forms enter into the concept of "being" as into that of "thought," but it never shows us *how* they are joined, nor *whence* they have their origin. Every question as to this origin, every reduction of the fundamental forms to an action of things or to a type of activity of mind would involve an obvious *petitio principii*: for the "whence" is itself nothing but a determinate form of logical relation. (1910/1923, p. 310)

Cassirer's talk of "a succession of logical acts" is not to be understood in terms of *psychological* acts of any particular cognitive subject, but refers rather to the succession of conceptual structures constituting the methodological progress of science. Any notion of cognitive subject must then emerge – by construction, as it were – within this process itself. And, once again, Cassirer's dissolution of "transcendental psychology" here closely parallels the Carnapian strategy of §66 of the *Aufbau*.

70 In this vein, too, the quotations from Cassirer in §II, this chapter, where he attempts to dissolve what he takes to be a false dichotomy between "idea" and "reality," also must be understood *in contrast* to the doctrines of the Southwest School. This is especially true of the final quotation (1910/1923, p. 242), where Cassirer explicitly contrasts "*form of being* [*Seinsform*]" with "*form of cognition* [*Erkenntnisform*]." I am indebted to Alan Richardson for emphasizing the importance of this point to me.

We have seen that Carnap, in the *Aufbau*, also represents empirical knowledge by a serial or stepwise methodological sequence. This sequence is intended to represent, not so much the historical series of mathematical-physical successor theories, but rather the epistemological progress of a representative individual – through which its knowledge extends from the initial subjective sensory data belonging to the autopsychological realm, through the world of public external objects constituting the physical realm, and finally to the intersubjective and cultural realities belonging to the heteropsychological realm. Carnap's methodological series is thus a "rational reconstruction" intended formally to represent the "actual process of cognition." For Carnap, as for Cassirer, we thereby represent the empirical side of knowledge by a methodological sequence of formal structures. For Carnap, however, this is not a sequence of successor theories ordered by the relation of relative refinement, but *a sequence of levels or ranks in the hierarchy of logical types of Principia Mathematica* – a sequence of levels *ordered by type-theoretic definitions.* Objects on any level (other than the first) are thus defined as classes of objects (or relations between objects) from the preceding level.[71]

Now, as suggested in §III, this chapter, Carnap's use of the type hierarchy of *Principia Mathematica* is indeed important for understanding his attempt to distance himself from the Marburg School in §5 of the *Aufbau.* Let me begin with the reminder that the title of §5 is "Concept and Object" and that its main thesis is that "the generality of a concept appears to us to be relative, and therefore the boundary between general concept and individual concept can be shifted in accordance with the point of view (§158)." It is following this conclusion, moreover, that Carnap situates his own notion of the "constitution" of objects with respect to the Marburg School (and I here reproduce the relevant passage once again):

> Are the constituted structures "generated in thought," as the Marburg School teaches, or "only recognized" by thought, as realism asserts? Constitutional theory employs a neutral language; according to it the structures are neither "generated" nor "recognized," but rather "*constituted*"; and it is already here to be expressly emphasized that this word "constitution" is always meant completely *neutrally*. From the point of view of constitutional theory the dispute involving "generation" versus "recognition" is therefore an idle linguistic dispute.

71 See especially the discussion of "ascension forms [*Stufenformen*]" in Part III.B of the *Aufbau*. There are exactly two such "ascension forms," namely, class and relation extensions (§40). As Carnap explains (1963a, p. 11), he first studied *Principia Mathematica* – whose type-theoretic conception of logic pervades the *Aufbau* – in 1919.

Carnap stresses the even more radical character of his conception immediately following:

> But we can go even further (without justifying it here) and say openly that a concept and its object are the same. This identity does not signify a substantialization of the concept, however, but, on the contrary, rather a "functionalization" of the object.

And the language of "substance" and "function" unmistakably suggests an allusion to Cassirer. It is not immediately clear, however, what Carnap has in mind here, but there is no doubt, as I have just argued, that the reading given in §III, this chapter, is quite mistaken. The key to a correct reading, I now believe, is to be found in §158.

What, for Carnap, is the relationship between a "concept" and "its object"? The first point to notice is that, as emphasized in §158, (almost) all structures in Carnap's constitutional system appear as classes (and classes of classes, etc.). All structures (except the elementary experiences at the very first level) therefore appear as higher-level objects in the hierarchy of types and are thus, in the terminology of §27, quasi objects. Hence, every structure (except those of the first level) appears both as a class (relative to objects of lower types) and as an element of classes in turn (relative to objects of higher types). And it is in this sense that (§5) "to every concept there belongs one and only one object, 'its object' (not to be confused with the objects falling *under* the concept."[72] Thus, the concept is the object viewed as a class (in relation to lower levels in the hierarchy of types), whereas the corresponding object is the concept viewed as an element of classes in turn (relative to higher levels). When Carnap states that a concept and its object are identical, therefore, he is simply saying that these two structures in the type hierarchy are in fact the very same structure.[73]

In the second place, however, as also explained in §158, the difference between "individual" objects belonging to the realm of reality and "universal" concepts having only an ideal being is, in the end, a purely formal-logical difference, which, accordingly, must itself be defined *within* the constitutional system. This difference in fact reduces to a formal-logical distinction between the spatiotemporal order and all other systems of order – ultimately, to the circumstance that there cannot be two different colors appearing at the same location in the visual field (whereas there can be two different locations

72 The point is reiterated at the very beginning of §158.

73 This point is correctly explained in §III, this chapter; the problem comes when I then emphasize §41 at the expense of §158.

with the same color). This formal-logical difference between the spatiotemporal order and all other systems of order (such as the order of colors) then makes the former system particularly well suited to play the role of a "principle of individuation" and thus a "principle of realization": objects belonging to the realm of reality are just those having determinate positions with respect to the *spatiotemporal* order.[74] Hence, although (almost) all objects – and, in particular, all real objects – are actually higher-level structures in the hierarchy of types, we can still articulate a purely formal-logical criterion for picking out the real objects from the ideal objects within this hierarchy.[75]

This conception of the individual or real object of empirical cognition stands in sharp contrast with that of the Marburg School. According to this latter conception, as we have seen, the real individual object of empirical cognition is never actually present in the methodological progress of science at all. On the contrary, we have only a sequence of ever more determinate structures, which, however, only becomes fully determinate and individualized in the ideal limit. Cassirer expresses the idea as follows:

> That this function [of empirical cognition] does not arrive at an end in any of its activities, that it sees rather behind every solution that may be given to it a new *task*, is certainly indubitable. Here, in fact, "individual [*individuelle*]" reality confirms its fundamental character of inexhaustibility. But, at the same time, this forms the characteristic advantage of true scientific relational concepts: that they undertake this task in spite of its incompletability in principle. Every new postulation, in so far as it connects itself with the preceding, constitutes a new step in the *determination* of being and happening. The individual [*Das Einzelne*] determines the direction of cognition as infinitely distant point. (1910/1923, p. 232)[76]

The real object of empirical cognition is thus the never-completed "X" toward which the methodological progress of science is converging.

Indeed, as explained above, this conception of the necessarily never fully realized object of empirical cognition is the essence of what the Marburg

74 Compare §172, where the concept of "real-typical [*wirklichkeitsartig*]" object is explained. Thus a "real-typical" object is an object *capable* of reality – as opposed to a "universal" concept having only ideal being. Section 172 refers to a work by B. Christiansen where the concept of real-typicality, defined as "empirical objectivity," is put forward as an interpretation of Kant's use of the term "object." (For the relevant formal-logical features of the visual field see §§88–91, 117–18.)

75 It follows, according to §52, that "all real objects (they are recognized in constitutional theory in the same sphere as real as in the factual sciences, see §170) are quasi-objects."

76 This passage comes at the very end of a section explicitly criticizing Rickert's view of conceptualization and individuation. This is the very section to which Carnap refers in §75 of the *Aufbau* (see footnote 28, this chapter).

School calls their "genetic" view of knowledge. Carnap, for his part, explicitly rejects this view:

> According to the conception of the *Marburg School* (cf. Natorp [Grundlagen] 18ff.) the object is the eternal X, its determination is an incompletable task. In opposition to this it is to be noted that finitely many determinations suffice for the constitution of the object – and thus for its univocal description among the objects in general. Once such a description is set up the object is no longer an X, but rather something univocally determined – whose complete description then certainly still remains an incompletable task. (§179)[77]

When Carnap, in §5, rejects the idea that the object of knowledge is "generated in thought," he is therefore rejecting precisely this "genetic" view of knowledge: the idea that the object of empirical knowledge, in contradistinction to the purely formal objects of mathematical knowledge, is to be conceived as a never-ending progression. Rather, for Carnap, all objects whatsoever – whether formal or empirical, ideal or real – are defined or "constituted" at *definite finite ranks* within the hierarchy of types, and, it is in *this* sense that Carnap's constitutional system is, in fact, entirely neutral between "realism" on the one side and "transcendental idealism" on the other.[78]

For the same reason, moreover, Carnap also rejects the idea of the *synthetic* a priori. Since an object is always defined or "constituted" at a definite finite rank, we can always separate those aspects of our characterization of the object that specify its definition (a purely structural definite description that picks it out uniquely from among all other objects) from those aspects that record further information about the object uncovered in the course of properly scientific investigation. Fully determining the latter is indeed an infinite task requiring the whole future progress of empirical science; establishing the former, however, is simply a matter of stipulation:

> After the first task, that of the constitution of the objects, [there] follows as the *second the task of investigating the remaining*, non-constitutional *properties* and relations of the objects. The first task is solved through a stipulation, the second, on the other hand, through *experience*. (According to the conception of constitutional theory there are no other components in cognition than

77 The reference is to Natorp (1910, Chapter 1, §§4–6). I am indebted to Alison Laywine for emphasizing to me the importance of this aspect of Natorp's view in the present connection.

78 It is in this sense, too, that Carnap is much more radical than Cassirer. For Carnap, there is no fundamental distinction between (ideal) mathematical and (real) empirical objects with respect to constructibility (as definite formal structures) within the constitutional system.

these two – the conventional and the empirical – and thus no synthetic a priori [components].) (§179)

And it is in this way that Carnap explicitly opposes the commitment to the synthetic a priori that is still maintained within the Marburg School.

Carnap thereby arrives at a radical transformation of the Marburg tradition. Since we have now definitively replaced the synthetic a priori with an analytic version, epistemology or philosophy is itself transformed into a logico-mathematical constructive project: the purely formal project of actually writing down the required purely structural definite descriptions within the logic of *Principia Mathematica.* This purely formal exercise is to serve as a *replacement* for traditional epistemology, in which we represent the "neutral basis" common to all traditional epistemological schools. In particular, all such schools are in agreement, according to Carnap, about the basis and subsequent elaboration of the actual process of cognition (see footnote 13, this chapter). Since the constitutional system precisely represents this common ground of agreement within the neutral and uncontroversial domain of formal logic itself, all "metaphysical" disputes between the competing schools are thereby finally and definitively dissolved. The fruitless disputes of the philosophical tradition (including those arising within the neo-Kantian tradition!) are replaced by an entirely uncontentious discipline based on the new mathematical logic, and philosophy (once again) becomes a science: for Carnap, a purely technical subject.[79]

It is therefore worth noting, finally, that the *Aufbau* construction suffers from serious technical problems – problems that, in fact, undermine Carnap's attempt to distinguish himself from the Marburg School with respect to the never-completed "X" of empirical cognition and the synthetic a priori. These problems are well known and arise precisely when Carnap attempts to move from the subjective realm of the autopsychological to the objective realm of the physical (see footnote 9, this chapter). Here, as we have seen, Carnap outlines a procedure for assigning colors to points of space-time ($\mathbf{R}^4$) in such a way that conditions of continuity and constancy are satisfied. It is never shown, however, that this assignment leads to a

79 See the Preface to the *Aufbau* (p. xvi): "The new type of philosophy has arisen in close contact with work in the special sciences, especially in mathematics and physics. This has the consequence that we strive to make the rigorous and responsible basic attitude of scientific researchers also the basic attitude of workers in philosophy, whereas the attitude of the old type of philosophers is more similar to a poetic [attitude].... the individual no longer undertakes to arrive at an entire structure of philosophy by a [single] bold stroke. Instead, each works in his specific place within the *single* total science."

definite stable object – that is, a definite relation between space-time points and colors – existing at a determinate rank in the type-theoretic hierarchy. Indeed, on closer consideration, such a definite stable object appears to be impossible: the assignment of colors (and, more generally, of "perceptual qualities") is continually and indefinitely revised as we progress through the hierarchy of types. This is because, first, the initial assignment – based on the "observations" of a single subject – is subsequently revised on the basis of both the reports of other subjects and the scientific regularities discovered in the world of physics; and second, the construction of the world of physics suffers from a precisely parallel ambiguity: the methodological procedure leading (via the "physico-qualitative coordination") from sensible qualities to numerical physical state-magnitudes also is continually revised as we progress through the hierarchy. The assignment of colors depends on the subsequently constructed assignment of physical state-magnitudes, and the latter depends on the former – and both depend, moreover, on the reports of other persons, which are themselves only available at a still later stage.[80]

Carnap's construction of the physical world therefore appears never to close off at a definite rank in the hierarchy of types: it is continually revised to infinity. And this means, of course, that the Marburg doctrine of the never completed "X" turns out to be correct – at least so far as physical (and hence all higher-level) objects are concerned. Whereas autopsychological objects receive actual definitions locating them at definite type-theoretic ranks (as sets of . . . sets of elementary experience), this is not and cannot be true for the higher-level objects. It follows, therefore, that Carnap's rejection to the synthetic a priori – according to which all characterizations of the objects of

80 For the continual – and mutual – revisability of the relevant assignments, see §§135, 136, 144. Note that this process is not a matter of our initial subject making revisions in the light of further experience, for all experience is assumed to be already in (§101); rather, the relevant revisions take place as the *totality* of given experience is successively reorganized at higher and higher levels in the hierarchy of types. Standard objections to Carnap's construction of the external world here (footnote 9, this chapter) do not bring out the full force of the problem, I think, for they simply amount to the observation that Carnap's methodological rules provide us with only an *implicit* definition of the desired assignment. This misses the point, for Carnap's method of purely structural definite description turns implicit definitions into explicit ones precisely by adding a *uniqueness* clause (see §15). If there is a definite and unique relation within the hierarchy of types between colors and space-time points (analogously, between sensory qualities and physical state-magnitudes) satisfying Carnap's methodological constraints, his procedure is therefore entirely unobjectionable. That such a definite and unique relation does not exist is suggested by the possibility of continual revision.

science are either definitions (conventional stipulations) or ordinary empirical truths (concerning already-constituted objects) – also fails. In particular, if the methodological principles (discussed in §I, this chapter) governing the relevant assignments at higher levels do not close off at a definite rank, they can no longer be viewed as simply definitional conventions: there are, in fact, no actual definitions to which they could (analytically) contribute. The status of these methodological principles thus remains fundamentally unclear, and Carnap ends up with no objection, in particular, to the synthetic a priori.[81] Indeed, Carnap is not able satisfactorily to address this problem until, in the 1930s, he fundamentally reconsiders both his conception of logic (analyticity) and his conception of epistemology.

81 Compare footnote 17 (this chapter), together with the text to which it is appended, and recall (footnote 19, this chapter) that Carnap explicitly views such methodological principles as synthetic a priori in his pre-*Aufbau* period. I am indebted to comments from Robert Nozick, and also from Alison Laywine (see footnote 77, this chapter), for helping me to clarify my thinking on this point.

PART THREE

LOGICO-MATHEMATICAL TRUTH

7

ANALYTIC TRUTH IN CARNAP'S *LOGICAL SYNTAX OF LANGUAGE*

Throughout his philosophical career, Carnap places the foundations of logic and mathematics at the center of his inquiries: he is concerned above all with the Kantian question "How is mathematics (both pure and applied) possible?" Although he changes his mind about many particular issues, Carnap never gives up his belief in the importance and centrality of this question, nor does he ever waver in his conviction that he has the answer: the possibility of mathematics and logic is to be explained by a sharp distinction between formal and factual, analytic and synthetic truth. Thus, throughout his career, Carnap calls for, and attempts to provide,

> an explication for the distinction between logical and descriptive signs and that between logical and factual truth, because it seems to me that without these distinctions a satisfactory methodological analysis of science is not possible. (1963b, p. 932)

For Carnap, it is this foundation for logic and mathematics that is distinctive of logical – as opposed to traditional – empiricism. As he puts it in his "Intellectual Autobiography" (1963a, p. 47): "It became possible for the first time to combine the basic tenet of empiricism with a satisfactory explanation of the nature of logic and mathematics." In particular, we can avoid the "non-empiricist" appeal to "pure intuition" or "pure reason" while simultaneously avoiding the naive and excessively empiricist position of J. S. Mill (*Ibid.*).

Indeed, from this point of view, Carnap's logicism and especially his debt to Frege become even more important than his empiricism and his connection with the Vienna Circle. The point has been put rather well, I think, by Beth in his insightful article in the Schilpp volume:

> His connection with the Vienna Circle is certainly characteristic of his way of thinking, but by no means did it determine his philosophy. It seems to me that the influence of Frege's teachings and published work has been much deeper. In fact, this influence must have been decisive, and the development of Carnap's ideas may be considered as characteristic of Frege's philosophy as well. (1963, pp. 470–1)

Carnap (1963b, pp. 927–8) endorses this assessment in his reply to Beth, and it is quite consistent with what he says about his debt to Frege elsewhere.[1]

Yet when one looks at *Logical Syntax*,[2] which is clearly Carnap's richest and most systematic discussion of these foundational questions, the idea that Carnap is continuing Frege's logicism appears to be quite problematic. Not only does Carnap put forward an extreme "formalistic" (purely syntactic) conception of the language of mathematics – a conception that, as he explicitly acknowledges, is derived from Hilbert and would be anathema to Frege (§84) – his actual construction of mathematical systems exhibits none of the characteristic features of logicism. No attempt is made to define the natural numbers: the numerals are simply introduced as primitive signs in both of Carnap's constructed languages. Similarly, no attempt is made to derive the principle of mathematical induction from underlying logical laws: in both systems it is introduced as a primitive axiom (in Language I it appears as a primitive schematic inference rule, R14 of §12). In short, Carnap's construction of mathematics is thoroughly axiomatic and, as he explicitly acknowledges (§84), appears to be much closer to Hilbert's formalism than to Frege's logicism.

Carnap's official view of this question is that he is putting forward a reconciliation of logicism and formalism, a combination of Frege and Hilbert that somehow captures the best of both positions (§84).[3] In light of the above, however, it must strike the reader as doubtful that anything important in Frege's position has been retained. For that matter, although Carnap employs formalist rhetoric and an explicitly axiomatic formulation of mathematics, nothing essential to Hilbert's foundational program appears to be retained either. Thus, no attempt is made to give a finitary consistency or conservativeness proof for classical mathematics. Carnap takes Gödel's incompleteness results to show that the possibility of such a proof is "at best very doubtful," and he puts forward a consistency proof in a meta-language essentially richer than classical mathematics (containing, in effect, classical

1 See, e.g., (1963a, pp. 4–6) and the Preface to the second edition of (1928a/1967, pp. v–vi).
2 Carnap (1934c/37); references are given parenthetically in the text by section numbers.
3 Compare also Beth (1963, pp. 475–82) and Carnap (1963b, p. 928).

mathematics plus a truth definition for classical mathematics), which, as Carnap again explicitly acknowledges (§§34h, 34i), is therefore of doubtful foundational significance. At this point, then, Carnap's claim to reconcile Frege and Hilbert appears hollow indeed. What he has actually done, it seems, is thrown away all that is most interesting and characteristic in both views.

Such an evaluation would be both premature and fundamentally unfair, however. To see why, we must look more closely at the centerpiece of Carnap's philosophy – his conception of *analytic truth* – and how that conception evolves from Frege's while incorporating post-Fregean advances in logic, in particular, advances due to Hilbert and Gödel.

The first point to bear in mind is the familiar one that Frege's construction of arithmetic is not simply the embedding of a special mathematical theory (arithmetic) in a more general one (set theory). Frege's *Begriffsschrift* is not intended to be a mathematical theory at all; rather, it is to function as the logical framework that governs all rational thinking (and therefore all particular theories) whatsoever. As such, it has no special subject matter (e.g., the universe of sets) with which we are acquainted by "intuition" or any other special faculty. The principles and theorems of the *Begriffsschrift* are implicit in the requirements of any coherent thinking about anything at all, and this is how Frege's construction of arithmetic within the *Begriffsschrift* is to provide an answer to Kant: arithmetic is in no sense dependent on our spatiotemporal intuition but is built into the most general conditions of thought itself. This, in the end, is the force of Frege's claim to have established the analyticity of arithmetic.

But why should we think that the principles of Frege's new logic delimit the most general conditions of all rational thinking? Wittgenstein's *Tractatus* attempts to provide an answer: this new logic is itself built into any system of representation we are willing to call a language. For, from Wittgenstein's point of view, the *Begriffsschrift* rests on two basic ideas: Frege's function/argument analysis of predication and quantification, and the iterative construction of complex expressions from simpler expressions via truth-functions. So any language in which we can discern both function/argument structure (in essence, where there are grammatical categories of intersubstitutable terms) and truth-functional iterative constructions will automatically contain all the logical forms and principles of the new logic as well. Since it is plausible to suppose that any system of representation lacking these two features cannot count as a language in any interesting sense, it makes perfectly good sense to view the new logic as delimiting the most general conditions of any rational thinking whatsoever. For the new logic is now seen as embodying the most general conditions of meaningfulness (meaningful representation) as such.[4]

Carnap enthusiastically endorses this Wittgensteinian interpretation of Frege's conception of analyticity, and he is quite explicit about his debt to Wittgenstein throughout *Logical Syntax* (§§14, 34a, 52) and throughout his career.[5] Yet, at the same time, Carnap radically transforms the conception of the *Tractatus*, and he does this by emphasizing themes that are only implicit in Wittgenstein's thought. It is here, in fact, that Carnap brings to bear the work of Hilbert and Gödel in a most decisive fashion.

First, Carnap interprets Wittgenstein's elucidations of the notions of language, logical truth, logical form, and so on as definitions in *formal syntax*. They are themselves formulated in a meta-language or "syntax language," and they concern the syntactic structure either of some particular object language or of languages in general:

> All questions of logic (taking this word in a very wide sense, but excluding all empirical and therewith all psychological reference) belong to syntax. *As soon as logic is formulated in an exact manner, it turns out to be nothing other than the syntax either of some particular language or of languages in general.* (§62)

This syntactic interpretation of logic is, of course, completely foreign to Wittgenstein. For Wittgenstein, there can be only one language – the single interconnected system of propositions within which everything that can be said must ultimately find a place; and there is no way to get "outside" this system so as to state or describe its logical structure: there can be no syntactic meta-language. Hence logic and all "formal concepts" must remain ineffable in the *Tractatus*.[6] Yet Carnap takes the work of Hilbert and especially Gödel to have decisively refuted these Wittgensteinian ideas (see especially §73). Syntax (and therefore logic) can be exactly formulated; and, in particular, if our object language contains primitive recursive arithmetic, the syntax of our language (and every other language) can be formulated within this language itself (§18).

Second, Carnap also clearly recognizes that the linguistic or "syntactic" conception of analyticity developed in the *Tractatus* is much too weak to embrace all of classical mathematics or all of Frege's *Begriffsschrift*. For the two devices of function/argument structure (substitution) and iterative truth-functional construction that were seen to underlie Frege's distinctive analysis

4 For the relationship between Frege and the *Tractatus*, see especially Ricketts (1985).

5 See especially (1963a, p. 25).

6 See *Tractatus*, 4.12–4.128. This is the basis for the "logocentric predicament" of Ricketts (1985). See also Goldfarb (1979).

of predication and quantification do not lead us to the rich higher-order principles of classical analysis and set theory. As Gödel's arithmetization of syntax again decisively shows, all that is forthcoming is primitive recursive arithmetic. Of course, the *Tractatus* is itself quite clear on the restricted scope of its conception of logic and mathematics in comparison with Frege's (and Russell's) conception. Wittgenstein's response to this difficulty is also all too clear: so much the worse for classical mathematics and set theory.[7]

Carnap's own response is quite different, however, for his aim throughout is not to replace or restrict classical mathematics but to provide it with a philosophical foundation: to answer the question "How is classical mathematics possible?" And it is here that Carnap makes his most original and fundamental philosophical move: we are to give up the "absolutist" conception of logical truth and analyticity common to Frege and the *Tractatus.* For Carnap, there is no such thing as *the* logical framework governing all rational thought. Many such frameworks, many such systems of what Carnap calls L-rules are possible; and all have an equal claim to "correctness." Thus, we can imagine a linguistic framework whose L-rules are just those of primitive recursive arithmetic itself (such as Carnap's Language I); a second whose L-rules are given by set theory or some higher-order logic (such as Carnap's Language II); a third whose L-rules are given by intuitionistic logic; a fourth whose L-rules include part of what is intuitively physics (such as physical geometry; compare §50); and so on. As long as the L-rules in question are clearly and precisely delimited within formal syntax, any such linguistic framework defines a perfectly legitimate language (*Principle of Tolerance*):

> *In logic there are no morals.* Everyone is at liberty to build up his own logic, i.e., his own form of language, as he wishes. All that is required of him is that, if he wishes to discuss it, he must state his methods clearly, and give syntactical rules instead of philosophical arguments. (§17)

Thus, Carnap's basic move is to relativize the "absolutist" and essentially Kantian program of Frege and the *Tractatus.*

Carnap's general strategy then is concretely executed as follows. First, within the class of all possible linguistic frameworks, one particular such framework stands out for special attention. A framework whose L-rules are

7 For the rejection of set theory, see *Tractatus,* 6.031. Frege's and Russell's impredicative definition of the ancestral is rejected at 4.1273; the axiom of reducibility is rejected at 6.1232–6.1233. Apparently, then, we are limited to (at most) predicative analysis – and even this may be too much because of the doubtful status of the axiom of infinity (5.535). The discussion of mathematics and logic at 6.2–6.241 strongly suggests a conception of mathematics limited to primitive recursive arithmetic.

just those of primitive recursive arithmetic has a relatively neutral and uncontroversial status – it is common to "Platonists," "intuitionists," and "constructivists" alike (§16); moreover, as Gödel's researches have shown, such a "minimal" framework is nonetheless adequate for formulating the logical syntax of any linguistic framework whatsoever – including its own. So this linguistic framework, Carnap's Language I, can serve as an appropriate beginning and "fixed point" for all subsequent syntactic investigation – including the investigation of much richer and more controversial frameworks.

One such richer framework is Carnap's Language II: a higher-order system of types over the natural numbers including (higher-order) principles of induction, extensionality, and choice (§30). This framework then will be adequate for much of classical mathematics and mathematical physics. Nevertheless, despite the strength of this framework, we can exactly describe its logical structure within logical syntax; and, in particular, we can show that the mathematical principles in question are *analytic-in-Language-II* – in Carnap's technical terminology, they are included in the L-rules ("logical" rules), not the P-rules ("physical" rules) of Language II. So we can thereby explain the "mathematical knowledge" of anyone who adopts (who speaks, as it were) Language II. Such knowledge is implicit in the linguistic framework definitive of meaningfulness for such a person, and it is therefore formal, not factual.

We are now in a position to appreciate the extent to which Carnap has, in fact, combined the insights of Frege and Hilbert and has, in an important sense, attempted a genuine reconciliation of logicism and formalism. From Frege (and Wittgenstein), Carnap takes the idea that the possibility of mathematics is to be explained by showing how its principles are implicit in the general conditions definitive of meaningfulness and rationality. Mathematics is built into the very structure of thought and language and is thereby forever distinguished from merely empirical truth. By relativizing the notion of logical truth, Carnap attempts to preserve this basic logicist insight in the face of all the well-known technical difficulties; and this is why questions of reducing mathematics to something else – to "logic" in some antecedently fixed sense – are no longer relevant. From Hilbert (and Gödel), Carnap takes the idea that primitive recursive arithmetic constitutes a privileged and relatively neutral "core" to mathematics and, moreover, that this neutral "core" can be used as a "metalogic" for investigating much richer and more controversial theories. The point, however, is not to provide consistency or conservativeness proofs for classical mathematics, but merely to delimit its logical structure: to show that the mathematical principles in question are analytic in a suitable language. Carnap hopes thereby to avoid the devastating impact of Gödel's incompleteness results.

Alas, however, it was not meant to be. For Gödel's results decisively undermine Carnap's program after all. To see this, we have to be a bit more explicit about the details of the program. For Carnap, a language or linguistic framework is syntactically specified by its formation and transformation rules, where these latter specify both axioms and rules of inference. The language in question then is characterized by its consequence relation, which is defined in familiar ways from the underlying transformation rules. Now, such a language or linguistic framework will contain both formal and empirical components, both "logical" and "physical" rules. Language II, for example, will contain not only classical mathematics but classical physics as well, including "physical" primitive terms (§40), such as a functor representing the electromagnetic field, and "physical" primitive axioms (§82), such as Maxwell's equations. The task of defining analytic-for-a-language, then, is to show how to distinguish these two components: in Carnap's technical terminology, to distinguish L-rules from P-rules, L-consequence from P-consequence (§§51, 52).

How is this distinction to be drawn? Carnap proceeds on the basis of a prior distinction between *logical* and *descriptive* expressions (§50). Intuitively, logical expressions include logical constants in the usual sense (connectives and quantifiers) plus primitive expressions of arithmetic (the numerals, successor, addition, multiplication, and so on). Given the distinction between logical and descriptive expressions, we then define the analytic (L-true) sentences of a language as those theorems (L- or P-consequences of the null set) that remain theorems under all possible substitutions of *descriptive* expressions (§51). In other words, what we might call "descriptive invariance" separates the L-consequences from the wider class of consequences *simpliciter*. But how is the distinction between logical and descriptive expressions itself to be drawn? Here Carnap appeals to the *determinacy* of logic and mathematics (§50): logical expressions are just those expressions such that every sentence built up from them alone is decided one way or another by the rules (L-rules or P-rules) of the language. That is, every sentence built up from logical expressions alone is provable or refutable on the basis of these rules. In the case of descriptive expressions, by contrast, although some sentences built up from them will no doubt be provable or refutable as well (e.g., in virtue of P-rules), this will not be true for all such sentences – for sentences ascribing particular values of the electromagnetic field to particular space-time points, for example. In this way, Carnap intends to capture the idea that logic and mathematics are thoroughly a priori.

It is precisely here, of course, that Gödelian complications arise. For, if our consequence relation is specified in terms of what Carnap calls *definite* syntactic concepts – that is, if this relation is recursively enumerable – then even the theorems of primitive recursive arithmetic (Language I) fail to be analytic; and the situation is even worse, of course, for full classical mathematics (Language II). Indeed, as we would now put it, the set of (Gödel numbers of) analytic sentences of classical first-order number theory is not even an arithmetical set, and so, it certainly cannot be specified by definite (recursive) means. Carnap himself is perfectly aware of these facts, and this is why he explicitly adds what he calls *indefinite* concepts to syntax (§45). In particular, he explicitly distinguishes (recursive and recursively enumerable) *d*-terms or rules of derivation from (in general nonarithmetical) *c*-terms or rules of consequence (§§47, 48).

Moreover, it is here that Carnap is compelled to supplement his "syntactic" methods with techniques we now associate with the name of Tarski: techniques we now call "semantic." In particular, the definition of analytic-in-Language-II is, in effect, a truth definition for classical mathematics (§§34a–d).[8] Thus, if we think of Language II as containing all types up to ω (all finite types), for example, our definition of analytic-in-Language-II will be formulated in a stronger meta-language containing quantification over arbitrary sets of type ω as well. In general, then, Carnap's definition of analyticity for a language of any order will require quantification over sets of still higher order. The extension of analytic-in-L for any L therefore will depend on how quantifiers in a meta-language essentially richer than L are interpreted; the interpretation of quantifiers in this meta-language can only be fixed in a still stronger language; and so on (§34d).[9]

But why should this circumstance cause any problems for Carnap? After all, he is quite clear about the technical situation; yet he nevertheless sees no difficulty whatever for his logicist program. It is explicitly granted that Gödel's results thereby undermine Hilbert's formalism; but why should they refute Frege's logicism as well? The logicist has no special commitment to the "constructive" or primitive recursive fragment of mathematics: he is quite happy to embrace all of classical mathematics. Indeed, Carnap, in his Principle of Tolerance, explicitly rejects all questions concerning the legitimacy or justification of classical mathematics. What the logicist wishes

8 For discussion of this point, see Coffa (1987).

9 Compare Beth (1963), who takes this situation to undermine both Carnap's logicism and his formalism – and, in fact, to require an appeal to some kind of "intuition." Carnap's ingenuous response (1963b, pp. 928–30) is most instructive.

to maintain is not a reduction or justification of classical mathematics via its "constructive" fragment (as Hilbert attempts in his finitary consistency and conservativeness proofs), but simply that classical mathematics is analytic: that it is true in virtue of language or meaning, not fact. So why should Gödel's work undermine *this* conception?[10]

To appreciate the full impact of Gödel's results here, it is necessary to become clearer on the fundamental differences between Carnap's conception of analyticity or logical truth and that of his logicist predecessors. For precisely these differences are obscured by the notion of truth-in-virtue-of-meaning or truth-in-virtue-of-language, especially as this notion is wielded by Quine in his polemic against Carnap. Thus, the early pages of "Two Dogmas of Empiricism" distinguish two classes of logical truths (Quine 1951/1963, pp. 22–3). A *general* logical truth – such as "No unmarried man is married" – is "a statement that is true and remains true under all reinterpretations of its components other than the logical particles." An *analytic* statement properly so-called – such as "No bachelor is unmarried" – arises from a general logical truth by substitution of synonyms for synonyms. This latter notion then is singled out for special criticism for relying on a problematic conception of meaning (synonymy); and this is the level on which Quine engages with Carnap.

The first point to notice is that *these* Quinean criticisms are indeed relevant to Carnap, but not at all to his logicist predecessors – for their analytic truths simply do not involve nonlogical constants in this sense. Thus, whereas Carnap's languages contain primitive arithmetical signs (the numerals, successor, addition, etc.), Frege's *Begriffsschrift* and Russell's *Principia* do not. In these systems, the arithmetical signs are, of course, defined via the logical notions of truth-functions and quantifiers (including quantifiers over higher types). So Carnap needs to maintain that arithmetical truths are in some sense true in virtue of the meanings of "plus" and "times," for example, whereas Frege and Russell do not – these latter signs simply do not occur in their systems.

But what about the logical notions that Frege and Russell assume: the notions we now call logical constants and Quine calls logical particles? Does the same problem not arise for them? Do we not have to assume that the general logical truths of the *Begriffsschrift* and *Principia* are true in virtue of the meanings of "and," "or," "not," "all," and "some"? According to Wittgenstein's

10 I am indebted to Thomas Ricketts and especially to Alberto Coffa for pressing me on this point. (Note added in 1998: I now think that the response presented here is not satisfactory. For further discussion, see Chapter 9, this volume, especially footnote 56.)

position in the *Tractatus*, the so-called logical constants do not, properly speaking, have meaning at all. They are not words like others for which a "theory of meaning" is either possible or necessary. Indeed, for Wittgenstein, "there are no . . . 'logical constants' (in Frege's and Russell's sense)"(5.4). Rather, for any language, with any vocabulary of "constants" or primitive signs whatsoever, there are the purely combinatorial possibilities of building complex expressions from simpler expressions and of substituting one expression for another within such a complex expression. These abstract combinatorial possibilities are all that the so-called logical constants express: "Whenever there is compositeness, argument and function are present, and where these are present, we already have all the logical constants" (5.47). Thus, for Wittgenstein, logical truths are not true in virtue of the meanings of particular words – whether of "and," "or," "not," or any others – but solely in virtue of "logical form," that is, the general combinatorial possibilities common to all languages regardless of their particular vocabularies.[11]

Now this conception – that logical truths are true in virtue of "logical form," and not in virtue of "meaning" in anything like Quine's sense – is essential to the antipsychologism of the *Tractatus*. For, if logic depends on the meanings of particular words – even "logical words" like "and," "not," and so on – then it rests, in the last analysis, on psychological facts about how these words are actually used. It then becomes possible to contest these alleged facts and to argue, for example, that a correct theory of meaning supports intuitionistic rather than classical logic, for example. For Wittgenstein, this debate, in these terms, simply does not make sense. Logic rests on no facts whatsoever, and certainly not on facts about the meanings or usages of English (or German) words. Rather, logic rests on the abstract combinatorial possibilities common to all languages as such. In this sense, logic is absolutely presuppositionless and thus absolutely uncontentious.

The problem for this Tractarian conception has nothing at all to do with the Quinean problem of truth-in-virtue-of-meaning or truth-in-virtue-of-language. Rather, the problem is that the logic realizing this conception is much too weak to accomplish the original aim of logicism: explaining how mathematics – classical mathematics – is possible. Frege's *Begriffsschrift* cannot provide the required realization because of the paradoxes; and neither can Russell's *Principia* because of the need for axioms like infinity and reducibility. The *Tractatus* itself ends up with a conception of logic that falls somewhere between truth-functional logic and a ramified type-theory without infinity or reducibility; and it ends up with a conception of mathematics apparently limited to primitive recursive arithmetic (see

11 For further discussion, see Chapter 8 (this volume).

footnote 7, this chapter). So, Wittgenstein indeed may have achieved a genuinely presuppositionless standpoint, but only by failing completely to engage the foundational question that originally motivated logicism.

At this point Carnap has an extremely ingenious idea. We retain Wittgenstein's purely combinatorial conception of logic, but it is implemented at the level of the meta-language and given an explicit subject matter: namely, the syntactic structure of any language whatsoever. At the same time, precisely because logic in this sense is implemented at the level of the meta-language not the object language, it no longer has the impoverishing and stultifying effect evident in the *Tractatus*. For, although our purely syntactic meta-language is to have a very weak, and therefore uncontroversial, underlying logic, we can nonetheless use it to describe – but not to justify or reduce – much stronger systems: in particular, classical mathematics. In this way Carnap hopes to engage, and in fact, to neutralize the basic foundational question. Logic, in the sense of logical syntax, can in no way adjudicate this question. Indeed, from Carnap's point of view, there is no substantive question to be adjudicated. Rather, logic in this sense constitutes a neutral metaperspective from which we can represent the consequences of adopting any and all of the standpoints in question: "Platonist," "constructivist," "intuitionist," and so on.

Corresponding to any one of these standpoints is a notion of logic (analyticity) in a second sense: a notion of analytic-in-L. Sentences analytic-in-L are not true in virtue of the abstract combinatorial possibilities definitive of languages in general, but in virtue of conventions governing this particular L – specifically, on those linguistic conventions that establish some words as "logical" and others as "descriptive." Hence it is at this point, and only at this point, that we arrive at the Quinean problem of truth-in-virtue-of-meaning. And it is at this point, then, that Carnap's logicism threatens to collapse into its dialectical opponent, namely, psychologism.

Carnap hopes to avoid such a collapse by rigorously enforcing the distinction between *pure* and *applied* (descriptive) syntax. The latter, to be sure, is an empirical discipline resting ultimately on psychological facts: it aims to determine whether and to what extent the speech dispositions of a given speaker or community realize or exemplify the rules of a given abstractly characterized language L – including especially those rules definitive of the notion of analytic-in-L. Pure syntax, on the other hand, is where we develop such abstract characterizations in the first place. We are concerned neither with the question of which linguistic framework is exemplified by a given community or speaker, nor with recommending one linguistic framework over others – classical over "constructive" mathematics, for example. Rather, our aim is to step back from all such questions and simply articulate the consequences of adopting any and all such frameworks. The propositions of pure syntax are therefore logical or analytic propositions in the first sense: propositions of the abstract, purely combinatorial metadiscipline of logical syntax.

Here is where Gödel's work strikes a fatal blow. For, as we have seen, Carnap's general notion of analytic-in-L is simply not definable in logical syntax so conceived, that is, conceived in the above "Wittgensteinian" fashion as concerned with the general combinatorial properties of any language whatsoever. *Analytic-in-L* fails to be captured in what Carnap calls the "*combinatorial analysis* ... of finite, discrete serial structures" (§2): that is, primitive recursive arithmetic. Hence the very notion that supports, and is indeed essential to, Carnap's logicism simply does not occur in pure syntax as he understands it. If this notion is to have any place at all, then, it can only be within the explicitly empirical and psychological discipline of applied syntax; and the dialectic leading to Quine's challenge is now irresistible.[12] In this sense, Gödel's results knock away the last slender reed on which Carnap's logicism (and antipsychologism) rests.

In the end, what is perhaps most striking about *Logical Syntax* is the way it combines a grasp of the technical situation that is truly remarkable in 1934 with a seemingly unaccountable blindness to the full implications of that situation. Later, under Tarski's direct influence, Carnap of course came to see that his definition of analytic-in-L is not a properly "syntactic" definition at all; and, in *Introduction to Semantics*,[13] he officially renounces the definitions of *Logical Syntax* (§39) and admits that no satisfactory delimitation of L-truth in "general semantics" is yet known (§§13, 16). Instead, he offers two tentative suggestions: either we can suppose that our meta-language contains a necessity operator, so that a sentence S is analytic-in-L just in case we have $N(S$ is true) in M^L; or we can suppose that we already have been given a distinction between logical and factual truth in our meta-language, so that a sentence S is analytic-in-L just in case "S is true" is analytic-in-M^L (§116).

From our present, post-Quinean vantage point, the triviality and circularity of these suggestions is painfully obvious, but it was never so for Carnap. He never lost his conviction that the notion of analytic truth, together with a fundamentally logicist conception of mathematics, stands firm and unshakable. And what this shows, finally, is that the Fregean roots of Carnap's philosophizing run deep indeed. Unfortunately, however, they have yet to issue in their intended fruit.

12 See Ricketts (1982).

13 Carnap (1942); parenthetical references in this paragraph are to section numbers of this work.

8

CARNAP AND WITTGENSTEIN'S *TRACTATUS*

In his "Intellectual Autobiography," Carnap makes it clear that Wittgenstein – that is, the Wittgenstein of the *Tractatus* – was the philosopher who, after Frege and Russell, had the greatest influence on Carnap's own philosophical thinking. For it was from Wittgenstein's *Tractatus* that he derived his characteristic conception of logical and analytic truth:

> For me personally, Wittgenstein was perhaps the philosopher who, besides Russell and Frege, had the greatest influence on my thinking. The most important insight I gained from his work was the conception that the truth of logical statements is based only on their logical structure and the meaning of the terms. Logical statements are true under all conceivable circumstances; thus their truth is independent of the contingent facts of the world. On the other hand, it follows that these statements do not say anything about the world and thus have no factual content. (1963a, p. 25)

Whereas Frege and Russell had shown that all mathematical truth is logical – and therefore, for those who accept the view that all logical truth is analytic, that mathematical truth is also analytic – Wittgenstein was the first to articulate the true nature of logical truth itself: the truths of logic are tautologies that necessarily hold in all possible circumstances and hence say nothing about the world.[1]

1 Carnap (1963a, p. 46) explains Wittgenstein's distinctive contribution to the articulation of the concept of analytic truth as follows: "I had learned from Frege that all mathematical concepts can be defined on the basis of the concepts of logic and that the theorems of mathematics can be deduced from the principles of logic. Thus the truths of mathematics are analytic in the general sense of truth based on logic alone. The mathematician Hans Hahn, one of the leading members of the [Vienna] Circle, had accepted the same conception under

This conception of the tautologous character of logical and mathematical truth represents, for Carnap, the most important point of agreement between his philosophy and that of the *Tractatus*. But there is also an equally important point of fundamental disagreement. Whereas the *Tractatus* associates its distinctive conception of logical truth with a radical division between what can be said and what can only be shown but not said – a division according to which logic itself is not properly an object of theoretical science at all – Carnap associates his conception of logical truth with the idea that logical analysis, what he calls "logical syntax," is a theoretical science in the strictest possible sense:

> Furthermore, there is a divergence on a more specific point which, however, was of great importance for our way of thinking in the Circle. We read in Wittgenstein's book that certain things show themselves but cannot be said; for example the logical structure of sentences and the relation between the language and the world. In opposition to this view, first tentatively, then more and more clearly, our conception developed that it is possible to talk meaningfully about language and about the relation between a sentence and the fact described. ... [I] pointed out that only the structural pattern, not the physical properties of the ink marks, were relevant for the function of language. Thus it is possible to construct a theory about language, namely the geometry of the written pattern. This idea led later to the theory which I called "logical syntax" of language. (1963a, p. 29)

Indeed, in *Logical Syntax of Language*, Carnap states his divergence from Wittgenstein here in particularly sharp and striking terms:

> Wittgenstein considers that the only difference between the sentences of the speculative metaphysician and those of his own and other researches into the logic of science is that the sentences of the logic of science – which he calls philosophical elucidations – in spite of their theoretical lack of sense, exert, practically, an important psychological influence upon the philosophical investigator, which the properly metaphysical sentences do not, or, at least, not in the same way. Thus there is only a difference of degree, and that a very vague one. The fact that Wittgenstein does not believe in the possibility of the exact formulation of the sentences of the logic of science has as its consequence

the influence of Whitehead and Russell's work, *Principia Mathematica*. Furthermore, Schlick, in his book *Allgemeine Erkenntnislehre* (1918), had clarified and emphasized the view that logical deduction cannot lead to new knowledge but only to an explication or transformation of the knowledge contained in the premises. Wittgenstein formulated this view in the more radical form that all logical truths are tautological, that is, that they hold necessarily in every possible case, therefore do not exclude any case, and do not say anything about the facts of the world."

> that he does not demand any scientific exactitude in his own formulations, and that he draws no sharp line of demarcation between the formulations of the logic of science and those of metaphysics. (1934c/1937, §73, pp. 283–4)

Carnap's divergence from Wittgenstein here is thus absolutely central to his own attempt to articulate a radically new conception of *scientific* and therefore nonmetaphysical philosophy.[2]

Finally, Carnap also explains that his conception of logical syntax – which enables him to break with Wittgenstein over the say/show distinction and the inexpressibility of logical analysis – is itself principally derived from metamathematical investigations of Hilbert, Tarski, and Gödel:

> My way of thinking was influenced chiefly by the investigations of Hilbert and Tarski in metamathematics, which I mentioned previously. I often talked with Gödel about these problems. In August 1930 he explained to me his new method of correlating numbers with signs and expressions. Thus a theory of the forms of expressions could be formulated with the help of concepts of arithmetic. He told me that, with the help of this method of arithmetization, he had proved that any formal system of arithmetic is incomplete and incompletable. When he published this result in 1931, it marked a turning point in the development of the foundation of mathematics.
>
> After thinking about these problems for several years, the whole theory of language structure and its possible applications in philosophy came to me like a vision during a sleepless night in January 1931, when I was ill. (1963a, p. 53)

What Carnap came to see in his fevered state was, as it were, that Wittgenstein's doctrine of analytic truth plus the metamathematical work of Hilbert, Tarski, and Gödel equals his own distinctive conception of the philosophical enterprise.

Wittgenstein's *Tractatus* therefore exerted a deep and basically two-pronged influence on Carnap's philosophy. On the positive side, Carnap derives the idea of the essentially nonfactual character of analytic – that is, logico-mathematical – truth directly from the *Tractatus.* On the negative side, by reacting against the Tractarian conception of the ineffability of logic, Carnap develops perhaps his most important philosophical contribution: the idea of philosophy as logical syntax of language and thus as itself an exact science. Finally, by putting these two sides together, Carnap arrives at the idea that philosophy, too, is essentially nonfactual and says nothing about the world:

2 It is not sufficiently appreciated, for example, that it is precisely through the conception of philosophy as exactly formulable logical syntax that Carnap replies to the standard objection according to which the theses of logical positivism are themselves unverifiable in terms of experience: see especially (1935a, pp. 36–8).

> The chief motivation for my development of the syntactical method, however, was the following. In our discussions in the Vienna Circle it had turned out that any attempt at formulating more precisely the philosophical problems in which we were interested ended up with problems of the logical analysis of language. Since in our view the issue in philosophical problems concerned the language, not the world, these problems should be formulated, not in the object-language, but in the meta-language. Therefore it seemed to me that the development of a suitable meta-language would essentially contribute toward greater clarity in the formulation of philosophical problems and greater fruitfulness in their discussions. (1963a, p. 55)[3]

In one way or another, then, virtually all of Carnap's most characteristic philosophical ideas and distinctions result from the influence – both positive and negative – of Wittgenstein's *Tractatus.*

I

In the *Aufbau,* Carnap's conception of logic is that of *Principia Mathematica,* and the influence of Wittgenstein's *Tractatus* here appears to be very slight. To be sure, Carnap uses the term "tautology":

> *Logic (including mathematics) consists only of conventional stipulations* about the use of signs *and of tautologies* on the basis of these stipulations. The signs of logic (and mathematics) therefore do not designate objects, but rather serve only for the symbolic fixing of these stipulations. (1928a/1967, §107, p. 178)

But Carnap does not in any way engage the issues actually involved in Wittgenstein's doctrine of tautology. For example, he raises no questions about the axioms of infinity, reducibility, and choice and simply takes it completely for granted that all the objects of classical mathematics are, in fact, generated by his "tautologies."[4]

3 Compare p. 64: The distinction between logical and factual truth leads also to a sharp boundary line between syntax as the theory of form alone, and semantics as the theory of meaning, and thus to the distinction between uninterpreted formal systems and their interpretations. These distinctions are meant not as assertions, but rather as proposals for the construction of a meta-language for the analysis of the language of science. In this way we obtain also a clear distinction between questions about contingent facts and questions about meaning relations. This difference seems to me philosophically important; answering questions of the first kind is not part of the philosopher's task, though he may be interested in analyzing them; but questions of the second kind lie often within the field of philosophy or applied logic.

4 As Carnap (1963a, p. 24) himself intimates, then, it seems that he did not really attempt to come to terms with the doctrines of the *Tractatus* until he moved to Vienna and participated in the discussions of the *Tractatus* within the Vienna Circle.

Carnap's first serious engagement with the Wittgensteinian concept of tautology occurs in his *Abriss der Logistik* (1929). He introduces the notion of a truth function in §3 along with an explicit reference to 4.442 of the *Tractatus.* The concept of tautology then is introduced via an example of one, described as follows:

> This function therefore has the peculiarity that it is *always true* no matter what truth values [*Aussagewerte*] pertain to the arguments. Such a function is called a "*tautology*." If *p* and *q* are determinate statements in a tautology, then the validity of the tautology is independent of whether *p* or *q* are true. The tautology therefore communicates nothing about the obtaining or not obtaining of the states of affairs designated through *p* and *q*; it is *contentless* [*inhaltsleer*] (but not senseless [*sinnlos*]). (1929, §4b)

Carnap emphasizes the crucial point that tautologies cannot be conceived simply as maximally or completely general truths, but rather are true solely in virtue of their logical form:

> The essential character of logical propositions, in contrast to empirical [propositions], does not lie in the circumstance that they are general assertions but rather in the circumstance that they are tautologies: they are necessarily true on the basis of their mere form, but also contentless. (§4d)

Accordingly, Carnap here counts neither the axiom of choice (§24b) nor the axiom of infinity (§24e) as logical principles. Moreover, following Ramsey, Carnap obviates the need for the axiom of reducibility by adopting the simple theory of types (§9b).

Carnap (1930b) provides a somewhat fuller discussion of the three problematic axioms of *Principia.* He introduces "the attempts at criticism and development" of logicism of Wittgenstein and Ramsey in §4. The key idea, once again, is that the problematic axioms fail to be tautologies:

> The tautologies are therefore always true, solely on the basis of their form, independently of whether the argument propositions are true or false. Every proposition of logic is a tautology; thus according to the conception of logicism also every proposition of mathematics. But the *axiom of infinity* is not valid on the basis of its form alone; it is valid, if at all, as it were accidentally. For many individual domains it is valid, for others not. Whether one can speak of an absolute individual domain appears problematic. The *axiom of choice* likewise does not hold on the basis of its form alone – in so far as under the "existence" of the choice class one does not understand the consistency of this assumption but rather its constructive exhibition; its validity depends on what propositional functions are taken to be given in the thought domain in question and which principles of construction for the derivation of

further propositional functions (and thereby classes) are set up. The *axiom of reducibility* can be shown not to be tautological. (§4)

So what is the status of these axioms? As above, the axiom of reducibility is dispensable within the simple theory of types on the basis of Ramsey's work. For the other two axioms, however, we must adopt Russell's expedient: they are not logical principles but special assumptions upon which certain key theorems of mathematics must be viewed as conditional. Strictly speaking, therefore, such theorems can no longer be viewed as parts of logic at all.

In this period, Carnap's conception of the tautologousness of logical truth thus corresponds rather well to that of the *Tractatus.* For Wittgenstein, like Carnap, begins by rejecting what he takes to be the conception of logical truth of Frege and Russell, namely, the conception of logical truths as completely universal or maximally general truths containing only variables and logical constants:

> The mark of the logical proposition is *not* general validity.
>
> For to be generally valid means *only*: to be accidentally valid for all things. An ungeneralized proposition can be just as tautologous as a generalized [proposition]. (6.1231)

And the crucial point, I take it, is that on the conception of logical truths as maximally general it is not at all clear that logical truths say nothing about the world. Indeed, on this conception it appears that logical truths are simply the most abstract and general truths about the world – as Russell famously and colorfully puts it (1919, Chapter XVI, p. 169): "logic is concerned with the real world just as truly as zoology, though with its more abstract and general features." If, however, to make a claim about the world is to assert some particular distributions of truth values to elementary propositions, and if tautologies are true for all possible such distributions of truth values, then it is manifest that logical truths make no claim whatsoever about the world. It is in *this* sense that logical truths are analytic (6.11).

It follows that the problematic axioms of *Principia Mathematica* – despite their maximally general form – cannot themselves be tautologies:

> One could call logical general validity essential, in contrast with accidental [general validity], for example, that of the proposition "all men are mortal." Propositions like Russell's "axiom of reducibility" are not logical propositions, and this explains our feeling that they, if true, could only be true through a happy accident. (6.1232)
>
> A world can be thought in which the axiom of reducibility is not valid. But it is clear that logic has nothing to do with the question whether our world is actually so or not. (6.1233)

At least to this extent, then, Carnap appears to be in close agreement with the conception of the *Tractatus*.[5]

Yet Carnap, unlike Wittgenstein, could not long remain satisfied with a conception of logico-mathematical truth according to which most of the truths of classical mathematics are not analytic after all. In this sense, Carnap attempts to remain faithful to Frege's and Russell's logicism nonetheless. Carnap describes the evolution of his position within the Vienna Circle as follows:

> [T]he purely logical character of some of the axioms used in the system of *Principia Mathematica* seemed problematic, namely, that of the axiom of reducibility, the axiom of infinity, and the axiom of choice. We were gratified to learn from the studies on the foundations of mathematics made by F. P. Ramsey that the so-called ramified theory of types used in the *Principia* is unnecessary, that a simple system of types is sufficient, and that therefore the axiom of reducibility can be dispensed with. With respect to the other two axioms we realized that either a way of interpreting them as analytic must be found or, if they are interpreted as non-analytic, they cannot be regarded as principles of mathematics. I was inclined towards analytic interpretations; but during my time in Vienna we did not achieve complete clarity on these questions. Later I came to the conviction that the axiom of choice is analytic, if we accept the concept of class which is used in classical mathematics in contrast to a narrower constructivist concept. Furthermore, I found several possible interpretations for the axiom of infinity, different from Russell's interpretation, of such a kind that they make this axiom analytic. The result is achieved, e.g., if not things but positions are taken as individuals. (1963a, pp. 47–8)

As Carnap indicates, his new understanding of the analyticity of such axioms as infinity and choice was only achieved after he left Vienna for Prague in 1931; it therefore belongs to his logical syntax period.

The key move of *Logical Syntax* is the rejection of the "logical absolutism" of the *Tractatus*. There is no longer a single language in which all meaningful

5 With respect to the axiom of infinity, however, it appears that Carnap's and Wittgenstein's views are not in fact in agreement. It appears that Wittgenstein himself does subscribe to the "absolute individual domain" that Carnap finds problematic; for Wittgenstein's notion of tautology – arrived at by considering all possible truth value assignments to a *given* set of elementary propositions – does not involve the consideration of variable "individual domains." Nevertheless, it appears that for Wittgenstein the axiom of infinity could nonetheless not be a tautology: if true it would rather show itself in language through the existence of an infinite number of distinct names (5.535). Wittgenstein does not explicitly discuss the axiom of choice, but it appears that he would be even more reluctant to count it as a logico-mathematical truth than is Carnap (6.031): "The theory of classes is entirely superfluous in mathematics. This hangs together with the circumstance that the generality we need in mathematics is not *accidental* [generality]."

sentences are formulated, and there is no longer a single set of privileged logical sentences (such as the tautologies of the *Tractatus*). Instead, there is an indefinite multiplicity of distinct formal languages or linguistic frameworks, each with its own characteristic set of logical truths or analytic sentences. Thus linguistic frameworks answering to a constructivist conception in which the axiom of choice, say, is not acceptable are perfectly permissible, but so too are linguistic frameworks answering to the classical conception of set – in which, therefore, the axiom of choice is a logical principle after all. The point is that there is simply no notion of logical "correctness" independent of a given formal language or linguistic framework; there is only the purely *internal* notion of correctness or validity relative to the specified logical rules of one or another such framework. Thus, for example, whether or not the axiom of infinity is derivable from some set of antecedently designated logical principles – as it is in Frege's original system – there can be no objection whatsoever to a formal language or linguistic framework in which the axiom of infinity is rather laid down as a primitive logical principle. For there is again simply no question – independently of one or another linguistic framework – whether the axiom of infinity is or is not a "correct" logical principle.

More specifically, both of the languages considered in *Logical Syntax* are "coordinate languages" in which the individual objects are designated by numerals and the individual quantifiers thus range over the natural numbers. These numerical coordinates are then viewed as possible "positions" for empirical objects. Sentences asserting the existence of (any number of) empirical objects are then formulated by introducing empirical descriptive function signs that take appropriate (empirically determined) values on some class of numerical coordinates or "positions." The axiom of infinity therefore asserts that there are an infinity of possible "positions" for empirical objects but leaves the number of empirical objects actually occupying these "positions" entirely undetermined: *this* number can be determined only by synthetic propositions. Yet it is an analytic proposition that there are an infinity of numerical coordinates or possible "positions" simply because this proposition is now taken as a primitive logical truth of the formal languages or linguistic frameworks in question.

At this point one might feel that all connection with *Wittgenstein's* notion of tautology and analyticity has been lost. How, in particular, can we continue to maintain that the axiom of infinity is true in all possible circumstances and thus says nothing about the facts of the world? Within each formal language or linguistic framework the logical signs or logical constants are sharply distinguished from the nonlogical or descriptive signs, and in the linguistic frameworks under consideration the axiom of infinity is true for all possible interpretations – more precisely, for all possible substitutions – of the primitive nonlogical or descriptive signs of the language. Of course there is no longer a single privileged set of logical constants – the primitive logical constants of *Principia Mathematica*, for example – but rather a distinct class of logical signs or logical constants characterizing each distinct formal language or linguistic framework. And in the two formal languages considered in *Logical Syntax* all the numerals happen to be logical signs (more precisely, zero and successor are designated by primitive logical constants). It follows that the axiom of infinity contains only logical signs or logical constants; and, since it is a primitive proposition in both formal languages in question, it is a theorem of these languages invariant under all substitutions of nonlogical or descriptive signs. Hence it is a logical or analytic truth in Carnap's sense within both linguistic frameworks.

II

Wittgenstein's doctrine of tautology is also based on a sharp distinction between all other meaningful signs and the so-called logical constants. I say "so-called" because the essence of Wittgenstein's view is that the logical constants are not meaningful signs at all – strictly speaking there are *no* logical constants:

> Here it is manifest that there do not exist "logical objects," "logical constants" (in Frege's and Russell's sense). (5.4)

Putting what I take to be the same point in a somewhat different way, Wittgenstein states as his "fundamental thought" that the so-called logical constants are not representative of, do not go proxy for, objects:

> The possibility of the proposition rests on the principle of the representation of objects by means of signs.
>
> My fundamental thought is that the "logical constants" are not representative. That the *logic* of the facts cannot be represented. (4.0312)

Thus Wittgenstein's conception of the logical constants is intimately connected with his picture theory of meaning on the one hand ("the principle of the representation of objects by means of signs"), and his idea that logical form is itself inexpressible on the other ("the *logic* of the facts cannot be represented").

Without attempting to enter into the manifold interpretive difficulties here, I will now try simply to touch on those aspects of Wittgenstein's conception most relevant to a comparison with Carnap. I will thus sketch, if you will, a Carnapian reading of the *Tractatus*.

All meaningful linguistic expression, in the *Tractatus*, rests on the elementary propositions. Elementary propositions consist of names, which, I assume, come in different logical categories – corresponding to individual signs, predicate and relation signs, function signs of higher logical types, and so on. Each such logical category is associated with a class of grammatically intersubstitutable signs. Elementary propositions – grammatically acceptable concatenations of names – then have two essential features: first, they themselves contain no logical constants (no quantifier signs or propositional connectives); second, they are all logically independent of one another. It follows that any arbitrary assignment of truth values to the elementary propositions depicts a possible complete description of a "world" delimited by these propositions and that any such complete description of a "world" is generated by an assignment of truth values to the elementary propositions. Thus, anything one might want to say about such a "world" can be said by picking out a set of truth-value assignments to elementary propositions.

All other propositions – all other meaningful claims about the world – are therefore truth functions of the elementary propositions: all other propositions are determined by picking out some subset of "rows" of the truth table whose arguments are the elementary propositions. The so-called logical constants then are simply expedients for expressing in practice how particular complex propositions – corresponding to particular subsets of truth-value assignments to elementary propositions (particular subsets of "rows" of the truth table whose arguments are the elementary propositions) – are iteratively generated from the elementary propositions via truth-functional operations. The propositional connectives generate complex propositions from finite classes of basis propositions; the quantifiers generate complex propositions from (potentially) infinite classes of basis propositions specified by permitting some particular constituent in a given proposition – some particular name – to vary arbitrarily within its grammatically admissible substitution class.

Strictly speaking, then, the so-called logical constants are entirely unnecessary for meaningful representation of the world: elementary propositions and truth-value assignments to elementary propositions completely exhaust the function of meaningful representation. This is why the 5.4s, which articulate Wittgenstein's new conception of the logical constants, are a commentary on 5: "The proposition is a truth function of the elementary propositions" – and follow 5.3:

> All propositions are results of truth-operations on the elementary propositions.
>
> The truth-operation is the mode and manner in which the truth-function arises out of the elementary proposition.
>
> ....
>
> Every proposition is the result of truth-operations on elementary propositions.

Accordingly, the "vanishing" of the so-called logical constants depends on the privileged and exhaustive role of the elementary propositions:

> This vanishing of the apparent logical constants also appears when "$\sim$ ($\exists$x). $\sim$ fx" says the same as "(x) . fx", or when "($\exists$x) . fx . x = a" says the same as "fa". (5.441)
>
> Indeed all logical operations are already contained in the elementary propositions. For "fa" says the same as "($\exists$x) . fx . x = a".
>
> Where there is compositeness, there is argument and function, and where there are the latter, there are already all the logical constants.
>
> One could say: *The one* logical constant is that which *all* propositions, according to their nature, have in common.
>
> But this is the general form of the proposition. (5.47)

In this sense, therefore, there are no such things as the logical constants: there is only the general form of the proposition, that is, the general form of a truth function (6), that is, the general circumstance that all meaningful assertions correspond to sets of truth-value assignments to elementary propositions.

Two consequences of fundamental importance now follow for the nature of logic. First, logic and logical form cannot themselves be the subject of meaningful representation, for logical forms are not objects that can be represented by signs at all. All meaningful representation is of the various truth possibilities of the elementary propositions, and, strictly speaking, there simply are no logical signs – no signs, as it were, representing logical form. Second, the so-called logical truths are tautologies: propositions that as truth functions come out true for all possible assignments of truth values to elementary propositions. Such tautological propositions therefore say nothing: they make no claim about the world for they leave *all* truth possibilities for the elementary propositions entirely open. Nevertheless, tautological propositions *show* something, namely, how various truth possibilities and truth operations (truth possibilities and truth operations corresponding to genuinely meaningful, factual propositions) are formally related. On this reading, then, the doctrine of tautology, the conception of the logical constants, the picture theory of meaning, and the say/show distinction indeed are intimately connected. They are all parts of an overarching conception based on the priority and exhaustiveness of the elementary propositions.

On this reading, finally, the fundamental technical problem with the Tractarian doctrine of tautology does not involve a limitation to truth-functional logic or the assumption that first-order quantificational logic is decidable. For, on the one hand, quantificational propositions are intended precisely to depict truth possibilities for (possibly) infinite sets of basis propositions, generated by varying via substitution some particular component name in a given basis proposition (5.2s); and, on the other hand, once one sees that quantification is thereby accommodated, there is no particular reason to impose a requirement of decidability: tautologies say nothing because they are consistent with *all* possible truth-value assignments to elementary propositions, not because they (supposedly) can be effectively calculated. The fundamental technical problem – from our Carnapian point of view – is rather that quantification is explained purely substitutionally. We are therefore unable to accommodate quantificational logic where a substitutional interpretation is inappropriate, and, in particular, we are unable to accommodate *higher-order* quantification.

Carnap himself explains this last problem as follows:

> Wittgenstein demonstrated this thesis [that all logical truths are tautologies] for molecular sentences (i.e., those without variables) and for those with individual variables. It was not clear whether he thought that the logically valid sentences with variables of higher levels, e.g., variables for classes, for classes of classes, etc., have the same tautological character. At any rate, he did not count the theorems of arithmetic, algebra, etc., among the tautologies. But to the members of the Circle there did not seem to be a fundamental difference between elementary logic and higher logic, including mathematics. Thus we arrived at the conception that all valid statements of mathematics are analytic in the specific sense that they hold in all possible cases and therefore do not have any factual content. (1963a, p. 47)

In *Logical Syntax*, as we know, Carnap uses "coordinate languages" in which the first-order variables range over the natural numbers. Therefore, a purely substitutional interpretation of the individual quantifiers is perfectly admissible here. However, in evaluating the higher-order quantifiers of Language II – a type-theoretical language sufficient for all of classical mathematics – Carnap came to see under Gödel's influence that a substitutional interpretation is inadequate (due to indefinable sets of numbers; §34c). The definition of analyticity for such a higher-order language can thus by no means proceed substitutionally: what we require, in effect, is a standard

Tarskian ("referential" or "objectual") truth definition for all of classical mathematics.[6]

Once again, therefore, Carnap's problem with the strict Tractarian conception of tautology rests on its inability to comprehend full classical mathematics. And it is for this reason that Carnap repeatedly insists that his own conception of logical or analytic truth, although it is self-consciously modeled on that of the *Tractatus*, has successfully captured the notion in question *for the first time*:

> In this way [viz. through Carnap's own work in *Logical Syntax* and later writings], the distinction between logical and factual truth, which had always been regarded in our discussions in the Vienna Circle as important and fundamental, was at last vindicated. In this distinction we had seen the way out of the difficulty which had prevented the older empiricism from giving a satisfactory account of the nature of logic and mathematics.... Our solution, based on Wittgenstein's conception, consisted in asserting the thesis of empiricism only for factual truth. By contrast, the truths in logic and mathematics are not in need of confirmation by observations, because they do not state anything about the world of facts, they hold for any possible combination of facts. (1963a, p. 64)

By extending Wittgenstein's conception of tautology in the manner of *Logical Syntax* – that is, by defining the logical truths as, in effect, those truths valid for all interpretations of the nonlogical or descriptive vocabulary, and, crucially, by allowing the numerals and the signs for ("referential" or "objectual") higher-order quantification to count as logical constants in an appropriate formal language – Carnap has, for the first time, explained why all the propositions of classical logic *and mathematics* are tautologous and say nothing about the world.

III

It is clear, however, that, by thus extending the notion of tautology, Carnap has also transformed this conception into something that would be completely unacceptable from the standpoint of the *Tractatus*. This can be seen most clearly by considering the radical differences between the two conceptions with respect to precisely the issue of the nature of the logical constants.

In the *Tractatus*, there is a single privileged set of so-called logical constants, namely, the classical truth-functional connectives together with

6 Again, for a discussion of this point, see Coffa (1987).

(substitutionally interpreted) classical quantifiers. This is because, once again, meaningful propositions are conceived initially as corresponding to subsets of truth possibilities for the elementary propositions, and logic is exhausted by the formal relations – basically relations of containment – between different such subsets of truth possibilities. Hence classical truth-functional logic, including (substitutionally interpreted) classical quantificational logic, is the only conceivable logical possibility. For Carnap, by contrast, there is no such thing as the unique "correct" logic and therefore no single privileged set of logical constants. To be sure, classical truth-functional logic – including classical first-order quantificational logic – is one possibility, and, for this language, Wittgenstein's list of the logical constants is indeed appropriate. But the important point, for Carnap, is that this is only one possibility among many others. Formal languages with other sets of logical constants are equally admissible and equally "correct." In particular, there can be no objection to extending the classical logical constants to include the identity sign, the numerals, and ("referential" or "objectual") higher-order quantification; and this, as we have seen, is how Carnap is able to hold that all of classical mathematics is analytic.

It follows that, when one considers a language suitable for classical mathematics, the essence of Carnap's position lies in a fundamental extension of the notion of logical constant far beyond the scope this notion had in Frege, in Russell, or especially in Wittgenstein. Indeed, in *Logical Syntax,* Carnap holds that the narrower notion of logical constant is not even well defined:

> Whether, in the construction of a system of the kind described, only logical symbols in the narrower sense are to be included amongst the primitive symbols (as by both Frege and Russell) or also the mathematical symbols (as by Hilbert), and whether only logical primitive sentences in the narrower sense are to be taken as [logically]-primitive sentences, or also mathematical sentences, is not a question of philosophical significance, but only one of technical expedience. In the construction of Languages I and II we have followed Hilbert and selected the second method. Incidentally, the question is not even accurately formulated; we have in general syntax made a formal distinction between logical and descriptive symbols, but a precise classification of the logical symbols in our sense into logical symbols in the narrower sense and mathematical symbols has so far not been given by anyone. (1934c/1937, §84, p. 327)

As we have seen, Carnap is perhaps a bit too casual here. For the only way in which he can continue to maintain that full classical mathematics is analytic is precisely by considering "Hilbertian" languages in which certain primitive mathematical symbols are now declared to be logical constants.

Yet in the *Tractatus*, of course, there can be absolutely no question of extending the class of logical constants to include primitive mathematical symbols. Indeed, as is well known, even the identity sign creates serious problems there, for identity cannot be conceived as a truth operation on elementary propositions (5.3s). The essence of the Tractarian conception lies rather in a contraction of the notion of logical constant that self-consciously assimilates this notion to that of "auxiliary" symbols such as brackets (5.461) and punctuation marks (5.4611) – that is, to signs that *manifestly* have no independent meaning. Carnap's tendency, on the contrary, is to move in precisely the opposite direction:

> Examples of signs which are regarded as **logical** are the sentential connectives ("∼", "∨", etc.), the sign of the universal operator ("for every"), the sign of the element-class relation ("∈", "is a"), auxiliary signs (e.g., parentheses and commas as ordinarily used in symbolic logic, punctuation marks in the written word languages), the sign of logical necessity in a (non-extensional) system of modalities ("N"). Further, all those signs are regarded as logical which are definable by those mentioned; hence e.g. the sign of the existential operator ("∃", or "for some"), signs for universal and null classes of all types, the sign of identity ("=", "the same as"), all signs of the system of [Princ. Math.] by Whitehead and Russell and of nearly all other systems of symbolic logic, all signs of mathematics ... with the meaning they have when applied in science, all logical modalities (e.g. Lewis' "strict implication"). (1942, pp. 57–8)

And Carnap elsewhere explains that parentheses, brackets, and the like are *mere* auxiliaries.[7] Thus, if we divide the putative candidates for logical constants into, first, brackets and other auxiliaries, second, logical signs in the traditional narrower sense, and third, primitive mathematical signs, the tendency of the *Tractatus* is to assimilate all logical constants to the first class, whereas Carnap's tendency is to assimilate them to the third class.

Accordingly, in *Logical Syntax* the logical constants are initially completely on a par with all other primitive symbols. All expressions are introduced purely syntactically, and no expression has any independently specified meaning. We begin simply with primitive expressions and strings thereof, and the logical expressions, in particular, are simply a syntactically specified subclass of expressions. Purely syntactic formation rules delimit the class of grammatical strings of expressions or sentences of the language in question; purely syntactic transformation rules delimit the class of theorems – that is,

7 See (1954/1958, p. 7): "Among the logical signs are the parentheses '(' and ')' and the comma ',' as in e.g. '$F(a, b)$'. However, these signs have only a subordinate role, analogous to that of punctuation marks."

consequences of the null class of sentences – of the language in question. In what Carnap calls general syntax, these transformation rules then induce a purely syntactic distinction between logical and descriptive signs (§50) and thus a purely syntactic distinction between logical and physical theorems (§51), analytic and synthetic truths (§52). All of these notions are therefore entirely relative to the transformation rules of a given formal language or linguistic framework, a circumstance that Carnap explicitly contrasts with Wittgenstein's "absolutism":

> Later, Wittgenstein made the same view [that analytic propositions are true solely on the basis of their formal structure] the basis of his whole philosophy. "It is the characteristic mark of logical sentences that one can recognize from the symbol alone that they are true; and this fact contains in itself the whole philosophy of logic" ([*Tractatus*] p. 156 [6.113]). Wittgenstein continues: "And so also it is one of the most important facts that the truth or falsehood of non-logical sentences can *not* be recognized from the sentences alone." This statement, expressive of Wittgenstein's absolutist conception of language, which leaves out the conventional factor in language-construction, is not correct. It is certainly possible to recognize from its form alone that a sentence is analytic; but only if the syntactical rules of the language are given. If these rules are given, however, then the truth or falsity of certain synthetic sentences – namely, the determinate ones – can also be recognized from their form alone. It is a matter of convention whether we formulate only [logical]-rules, or include [physical]-rules as well; and the [physical]-rules can be formulated in just as strictly formal a way as the [logical]-rules. (1934c/1937, §52, p. 186)

Needless to say, a language in which certain primitive laws of physics are built into its logical syntax at the outset would be completely anathema from the point of view of the *Tractatus.*

In sum, we might express the essential difference between the two conceptions as follows. The *Tractatus* begins with a philosophical picture of the nature of meaningful representation. All meaningful claims correspond to subsets of truth possibilities for a given set of elementary propositions – which are entirely independent of one another so that all possible distributions of truth-value assignments are initially possible. Furthermore, the elementary propositions are composite – they consist of names that can be independently varied within given substitution classes – but they nonetheless contain no logical constants. Wittgenstein's conception of the logical constants as truth operations on the elementary propositions and his conception of analytic propositions or tautologies as holding for every possible distribution of truth values to the elementary propositions then follow directly from his initial philosophical picture of meaningful representation. Carnap, by contrast, begins with no such philosophical picture. He begins simply with the idea of a syntactically specified formal language given by formation rules and transformations rules. There is no independent notion, in particular, of what it means for a sign to represent an object in the world or for a sentence to make an assertion or claim about the world. Carnap instead exploits syntactic features of the transformation rules to define a distinction between logical and descriptive signs, and this distinction then induces a distinction between analytic and synthetic sentence – which, as it were, *syntactically represents* the notion of meaningful claim about the world. Carnap's assertion that analytic sentences are empty of factual content and make no real claim about the world has therefore an entirely different sense and force from Wittgenstein's similar-sounding assertion.

IV

By the same token, Carnap's assertion that logical form and logical syntax are perfectly capable of exact expression has very little to do with the Tractarian denial of a similar-sounding proposition. What Carnap means is that if we view language purely syntactically (as consisting simply of arbitrary expressions and strings thereof), then we can perfectly well describe the logical syntax of such a language – that is, its formation and transformation rules – in a syntactic meta-language. For example, from the perspective of a syntactic meta-language the grammatical sentences are a certain subset of the set of all possible strings of primitive symbols, and the transformation rules depict certain relations between such strings of symbols. Further, and what is of most relevance to our central theme, the logical constants are simply a syntactically specified subclass of expressions defined in a particular way from the transformation rules. Logical syntax is thus the theory of a certain class of formal structures – combinatorial structures generated by strings of symbols – and it is in this sense that Carnap (1934c/1937, §73, p. 283) asserts that "*syntax is exactly formulable in the same way as geometry is.*"

By contrast, when the *Tractatus* denies that logical syntax is expressible, it of course does not intend to deny that one can formulate a combinatorial theory of strings of symbols. But, from the point of view of the *Tractatus*, this would simply be a particular theory formulated within a more comprehensive language – a language that embodies and presupposes logical form and logical syntax in precisely the sense of the *Tractatus*. The symbol in Carnap's syntactic meta-language designating the negation sign in the object language, for example, has nothing to do with the truth operation associated with negation in the sense of the *Tractatus.* On the contrary, Carnap's symbol is simply a *name* for a particular object – of the negation symbol in the object language. The truth operation associated with negation therefore is not represented by this symbol. Indeed, according to the *Tractatus,* this operation is, of course, not *represented* at all; it is rather *shown* by the use of whatever symbol expresses negation in the meta-language. And Carnap's syntactic meta-language now becomes *the* language in Wittgenstein's sense – that is, *the* language in which all logical relations are expressed:

> The proposition can present the whole of reality, but it cannot present what it must have in common with reality in order to be able to present it – logical form.
>
> In order to be able to present logical form we would have to be able to set ourselves up outside of logic with the proposition, and this means outside the world. (4.12)

Language in the *Tractatus,* that is, *the* language, is that system of symbols through which all truth possibilities for the elementary propositions are presented; and logical form, that is, the general form of the proposition, cannot itself be presented because truth operations on elementary propositions are not, in *this* language, a possible object of representation via names. A mere combinatorial structure of strings of expressions is not a *language* in the sense of the *Tractatus.*

Now Carnap, in *Logical Syntax,* is sensitive to the circumstance that in the *Tractatus* there is only a single language and that, accordingly, the idea that the logical syntax of *this* language could be expressed in a syntactic meta-language is completely unacceptable from the Tractarian point of view. And it is precisely here, in fact, that Carnap brings to bear the method of arithmetization introduced by Gödel:

> Up to the present, we have differentiated between the object-language and the syntax-language in which the syntax of the object-language is formulated. Are these necessarily two separate languages? If this question is answered in the affirmative (as it is by Herbrand in connection with metamathematics), then a third language will be necessary for the formulation of the syntax of the syntax-language, and so on to infinity. According to another opinion (that of Wittgenstein), there exists only one language, and what we call syntax cannot be expressed at all – it can only "be shown." As opposed to these views, we intend to show that, actually, it is possible to manage with one language only; not, however, by renouncing syntax, but by demonstrating that without the emergence of any contradictions the syntax of this language can be formulated within this language itself. In every language S, the syntax of any

> language whatsoever – whether of an entirely different kind of language, or of a sub-language, or even of S itself – can be formulated to an extent which is limited only by the richness of the means of expression of the language S. (1934c/1937, §18, p. 53)

In his following section (§19), Carnap outlines Gödel's arithmetization of syntax. Carnap's idea, then, is that one can respond to the Tractarian doctrine that there is only a single language by, first, conceiving this language as itself a purely syntactic object, and, second, using Gödel's method of arithmetization to correlate numbers with symbols. The syntactic meta-language is thereby embedded in the object language (provided that the object language includes elementary arithmetic), and, in this precise sense, the object language can express its own syntax.

Yet it is clear once again, I think, that this ingenious suggestion does not in fact address the concerns of the *Tractatus* at all. For, once again, the *Tractatus* would emphatically reject Carnap's very first step: the conception of language as a purely combinatorial syntactic object. Thus, just as the *Tractatus* would insist that the expression designating the object-language symbol for negation in a syntactic meta-language has nothing to do with real negation (i.e., with this particular truth operation on elementary propositions), it would for the same reason insist that the number correlated with the negation symbol (in Carnap's arithmetization, the number 21) has nothing to do with real negation either. Real negation can be expressed neither by a meta-linguistic name nor by a numeral; it is rather *shown* by the use of whatever symbol is associated with that particular truth operation on elementary propositions.

Carnap, characteristically, has transformed an originally philosophical point into a purely technical question – in this case, the technical question of what formal theories can or cannot be embedded in a given object language. Considered purely as a technical question, however, the situation turns out to be far more complicated than it initially appears; and this circumstance is, I think, not without interest from the point of view of our present concerns. For it turns out, again as a consequence of Gödel's work, that it is as a matter of fact not possible in most cases of interest to express the logical syntax of a language in Carnap's sense in the language itself. Indeed, the problem arises precisely in connection with syntactically representing the crucial distinction between logical and descriptive constants.

I mentioned several times above that, in *Logical Syntax*, a distinction between logical and descriptive signs is induced by the transformation rules of the language in question (§50). Given this distinction, a division of all theorems of the language into logical and physical, analytic and synthetic, then follows easily: logical or analytic theorems remain theorems under all substitutions of descriptive expressions (§§51, 52).[8] How then are logical signs distinguished from descriptive signs? The idea is that the logical signs are those signs such that *all* sentences containing only these signs are determinate in the language – that is, either they or their negations are theorems of the language. For sentences essentially containing descriptive signs, by contrast, although *some* such sentences may be theorems of the language as well (e.g., general laws of physics), this will not be the case for *all* such sentences (e.g., sentences ascribing particular values at particular space-time coordinates to various physical fields). In this way, Carnap accommodates the pretheoretical idea that the meaning and application of logical signs is entirely determined by the syntactical rules of the language in question.

It is immediate from Gödel's work, however (and Carnap himself is completely clear about this point), that, for a language containing classical arithmetic, the distinction between logical and descriptive signs – and thus the distinction between analytic and synthetic sentences – can be correctly drawn only if the transformation rules of the language are nonrecursive. Indeed, it immediately follows that the notion of theoremhood for such a language must be nonarithmetical and thus not definable in the language itself. The distinction between logical and descriptive signs and the corresponding notion of analytic sentence can only be defined in a language containing classical mathematics plus a (Tarskian) truth definition for classical mathematics and thus in a language essentially richer than the language with which we began. Thus, the logical syntax in Carnap's sense for a language for classical mathematics can be expressed only in a distinct and essentially richer meta-language; the logical syntax for this meta-language can itself be expressed only in a distinct and essentially richer meta-meta-language; and so on. And it is precisely this situation that Carnap has in mind when he qualifies his introduction of the method of arithmetization in §18 (cited earlier) by the phrase "*to an extent which is limited only by the richness of the means of expression of the language S.*"

In commenting upon this consequence of Gödel's incompleteness theorem – namely, the circumstance that an infinite sequence of richer and richer meta-languages thus is necessary after all – Carnap remarks:

8 In *Introduction to Semantics* this same idea is given a semantical reading: the logical or analytic truths remain true for all *interpretations* of the descriptive predicates: (1942, §16, 2a). As Carnap states, this change accommodates Tarski's work and, in particular, the existence of indefinable properties.

> This is the kernel of truth in the assertion made by Brouwer [*Sprache*], and, following him, by Heyting [*Logik*], p. 3, that mathematics cannot be completely formalized. In other words, *everything mathematical can be formalized, but mathematics cannot be exhausted by one system*; it requires an infinite series of ever richer languages. (1934c/1937, §60d, p. 222)

Does this same situation not represent the kernel of truth – from Carnap's point of view, of course – in Wittgenstein's doctrine of the inexpressibility of logical syntax?

9

TOLERANCE AND ANALYTICITY IN CARNAP'S PHILOSOPHY OF MATHEMATICS

In *The Logical Syntax of Language,*[1] Carnap attempts to come to terms philosophically with the debate in the foundations of logic and mathematics that raged throughout the 1920s – the debate, that is, between the three foundational schools of logicism, formalism, and intuitionism. Carnap himself, as a student of Frege, Russell, and Wittgenstein, is of course most sympathetic to the logicist school. Nevertheless, he recognizes that traditional logicism cannot succeed: we cannot reduce mathematics to logic in some antecedently understood sense, whether in the sense of Frege's *Begriffsschrift* or Whitehead and Russell's *Principia Mathematica.* And, at the same time, Carnap is sensitive to the contributions, both technical and philosophical, of the two competing schools. In particular, he is sensitive to the notion of constructibility emphasized by the intuitionist school and, especially, to Hilbert's conception of *metamathematics* emphasized by the formalist school. Indeed, Carnap begins *Logical Syntax* by explaining that the meta-language, where we speak about the formulas and rules of a logical system, represents "what is essential in logic" (Foreword). The point of Carnap's book is then to develop a precise and exact method, logical syntax, wherein these "sentences about sentences" can be formulated.

Nevertheless, Carnap by no means shares Hilbert's foundational program. Carnap is fully cognizant, in particular, of Gödel's recently discovered

1 (1934c/1937); I cite this work in the text parenthetically by section numbers (and, in some few cases, by page numbers of the English translation followed in brackets by those of the German original). As Carnap explains in the Preface to the translation, the English version contains some important sections that were written for the original but not included because of lack of space. These sections appear in the translation with lowercase letters appended to the original numbering.

incompleteness theorems and accordingly states explicitly that "whether ... Hilbert's aim can be achieved at all, must be regarded as at best very doubtful in view of Gödel's researches on the subject" (§34i). Carnap thus shows no interest whatever in the foundational project of proving the consistency of classical mathematics within an essentially weaker, finitary, meta-language. Moreover, it is also clear that Carnap does not share the foundational concerns of traditional intuitionism or constructivism. To be sure, Carnap devotes a considerable portion of *Logical Syntax* (Parts I and II) to the articulation and investigation of a logical system, Language I, in which "[s]ome of the tendencies which are commonly designated as 'finitist' or 'constructivist' find, in a certain sense, their realization" (§16). Indeed, Carnap originally had felt substantial sympathy with the constructivism of Brouwer, Heyting, and Weyl.[2] Nevertheless, although, in harmony with these sympathies, Carnap at first had intended to develop Language I alone, he soon adopted the standpoint of "tolerance" according to which a language containing all of classical arithmetic, analysis, and set theory – Language II – is equally possible and legitimate.[3] In this sense, Carnap abandons all constructivist philosophy in *Logical Syntax* and instead views his particular constructivist system, Language I, as simply one possible formal-logical system among an infinity of equally possible such systems.

Carnap, in fact, intends to represent the most general possible logico-mathematical pluralism in *Logical Syntax.* We are entirely free to set up any system of formal rules we like, whether or not these rules represent the point of view of logicism, formalism, intuitionism, or any foundational school at all. And we are entirely free here because there can be no question of "justification" or "correctness" *antecedent* to the choice of one or another formal-logical system. From the point of view of the meta-language – that is, from the point of view of logical syntax – our task is neither to justify nor

2 See Carnap (1963a, p. 49): "[T]he constructivist and finitist tendencies of Brouwer's thinking appealed to us [viz. the Vienna Circle] greatly.... I had a strong inclination toward a constructivist conception." Here Carnap is referring to the period surrounding Brouwer's famous lecture to the Circle on intuitionism (which apparently also greatly influenced Wittgenstein) in March 1928. For discussion, see Menger (1994, pp. 130–9). Menger, as Carnap notes in the same passage, had in turn studied with Brouwer in Amsterdam during 1925–6.

3 See Carnap (1963a, pp. 55–6): "Originally, in agreement with the finitist ideas with which we sympathized in the Circle, I had the intention of constructing only language I. But later, guided by my own principle of tolerance, it seemed desirable to me to develop also the language form II as a model of classical mathematics. It appeared more fruitful to develop both languages than to declare the first language to be the only correct one or to enter into a controversy about which of the two languages is preferable." As Carnap notes in *Logical Syntax* (§17), he was influenced here especially by Menger (1930/1979).

to criticize any particular choice of rules, but rather to investigate and to compare the consequences of any and all such choices:

> From this point of view the dispute between the different tendencies in the foundations of mathematics also disappears. One can set up the language in its mathematical part as one of the tendencies prefers or as the other prefers. There is no question of "justification" here, but only the question of the syntactic consequences to which one or another choice leads – including also the question of consistency. (Foreword)

As Carnap here intimates, once we have made the choice of a particular formal-logical system, there is then a specific notion of logical "correctness" fixed by the rules in question, a notion of logical "correctness" *relative to* the formal rules (and their syntactic consequences) to which we have committed ourselves. For the choice of one such formal system over another, however, there is and can be no notion of "correctness." Here we are faced with a purely pragmatic or conventional question of suitability and/or convenience relative to one or another given purpose. If one is especially concerned to avoid the threat of inconsistency, for example, the choice of a relatively weak constructivist or intuitionist language such as Carnap's Language I is prudent. If, however, one wants the full power of classical mathematics (perhaps in view of ease of physical application), then one has no choice but to adopt a much richer language such as Carnap's Language II.[4] As Carnap expresses the resulting *principle of tolerance*: "*we do not wish to set up prohibitions, but rather to stipulate conventions*" (§17).

I

We have seen that Gödel's incompleteness results form an essential part of the background to *Logical Syntax*. Indeed, Carnap himself ascribes a central role in the genesis of *Logical Syntax* to his own interactions with Gödel in connection with these fundamental discoveries:

> [T]he members of the Circle, in contrast to Wittgenstein, came to the conclusion that it is possible to speak about language and, in particular, about the structures of linguistic expressions. On the basis of this conception, I

4 See Carnap (1963a, p. 49): "It is true that certain procedures, e.g., those admitted by constructivism or intuitionism, are safer than others. Therefore, it is advisable to apply these procedures as far as possible. However, there are other forms and methods which, though less safe because we do not have a proof of their consistency, appear to be practically indispensable for physics. In such a case there seems to be no good reason for prohibiting these procedures as long as no contradiction has been found." Compare Carnap (1939, §20).

> developed the idea of the logical syntax of a language as the purely analytic theory of the structure of its expressions. My way of thinking was influenced chiefly by the investigations of Hilbert and Tarski in metamathematics .... I often talked with Gödel about these problems. In August 1930 he explained to me his new method of correlating numbers with signs and expressions. Thus a theory of the forms of expressions could be formulated with the help of the concepts of arithmetic. He told me that, with the help of this method of arithmetization, he had proved that any formal system of arithmetic is incomplete and incompletable. When he published this result in 1931, it marked a turning point in the development of the foundations of mathematics.
>
> After thinking about these problems for several years, the whole theory of language structure and its possible applications in philosophy came to me like a vision during a sleepless night in January 1931, when I was ill. On the following day, still in bed with a fever, I wrote down my ideas on forty-four pages under the title, "Attempt at a metalogic." These shorthand notes were the first version of my book *Logical Syntax of Language*. In the spring of 1931 I changed the form of language usage dealt with in this essay to that of a coordinate language of about the same form as that later called "language I" in my book. Thus arithmetic could be formulated in this language, and by use of Gödel's method, even the metalogic of the language could be arithmetized and formulated in the language itself. (1963a, pp. 53–4)[5]

Given the central importance, for Carnap, of Wittgenstein's more general conception of logic and logical syntax, it is crucial for him to reject Wittgenstein's doctrine of the inexpressibility of logical syntax in favor of his own project of developing an explicit formal theory thereof.[6] And it is no wonder, then, that Carnap took Gödel's discoveries to be so important.

Yet Gödel himself, in a contribution written for inclusion in the Schilpp volume but never published, argues that the incompleteness results are incompatible with Carnap's position in *Logical Syntax*.[7] Gödel argues that, if the choice of logico-mathematical rules is really to be viewed as conventional, then we must have independent assurance that these rules do not have unintended empirical or factual consequences. We must know, that is, that the rules in question are conservative over the purely conventional realm. We therefore need to show that the rules are consistent, and this,

5 See §18 of *Logical Syntax*, entitled "The syntax of [language] I can be formulated in [language] I," where Carnap explains how Gödel's arithmetization of syntax (§19) allows us to overcome Wittgensteinian scruples.

6 I discuss both Carnap's debt to Wittgenstein and their divergence over the expressibility of logical syntax in Chapter 8 (this volume).

7 Gödel preserved six drafts of this contribution (entitled "Is Mathematics Syntax of Language?"), two of which appear in Gödel (1995).

by Gödel's second theorem, cannot be done without using a meta-language whose logico-mathematical rules are themselves even stronger than those whose conservativeness is in question. Hence, we can have no justification for considering mathematics to be purely conventional, for an unintended incursion into the empirical or factual realm cannot be excluded without vicious circularity. Gödel takes this state of affairs to support his own view, in opposition to Carnap's logicism, that there is no distinction in principle between the mathematical and the empirical or factual sciences: both deal with realms of objects given to us by intuition (rational and sensible, respectively).

Thomas Ricketts and Warren Goldfarb have recently contributed subtle and perceptive discussions of *Logical Syntax*, which seek, among other things, to defend Carnap against Gödel's challenge.[8] They point out, in particular, that Gödel's argument proceeds from the assumption that we are given antecedently a clear notion of the factual or empirical realm. Carnap then is depicted, accordingly, as starting with an unproblematic realm of empirical facts to which the logico-mathematical sentences are to be conventionally or stipulatively added.[9] Ricketts and Goldfarb rightly emphasize that such a language-independent notion of the factual or empirical realm is foreign to Carnap himself. Instead, Carnap holds that the very distinction between the conventional and the factual itself only makes sense relative to, or within, a given formal language: the conventional statements relative to a given formal-logical system are just the sentences that are *analytic* relative to this system, the empirical statements relative to a given formal-logical system are just those sentences that are *synthetic* relative to this system.[10] To be given a formal language is thus to be given at the same time a distinction between analytic and synthetic (conventional and factual) sentences, and there can therefore be no further question of showing that the logico-mathematical

8 See Ricketts (1994), Goldfarb and Ricketts (1992). See also Goldfarb (1995).

9 Thus Gödel (1995, p. 335, n. 9) begins the third draft of "Is Mathematics Syntax of Language?" by citing the following passage from the conclusion of Carnap (1935b/1953, p. 128): "When formal science is added to factual science [*Realwissenschaft*] *no new objects* are thereby introduced, as many philosophers who oppose the 'formal' or 'spiritual [*geistig*]' or 'ideal' objects to the 'real [*real*]' objects of factual science believe. *Formal science has no objects at all*; it is a system of auxiliary sentences, free of objects and empty of content."

10 Carnap explains the dependence of the distinction between formal and factual science on his analytic/synthetic distinction in "Formalwissenschaft and Realwissenschaft" itself – which here simply follows *Logical Syntax*. From this point of view, as Goldfarb and Ricketts rightly emphasize, statements like that cited by Gödel in footnote 9, this chapter, are simply colorful formulations, in the material mode, of the idea that logico-mathematical sentences are analytic.

or analytic sentences of the language do not turn out, inadvertently as it were, to include factual sentences.[11]

Carnap presents a formal explication of the distinction between analytic and synthetic sentences in §§50–2 of *Logical Syntax*. This explication is carried out within what Carnap calls "general syntax," and it is thus meant to apply to arbitrary formal languages. It is meant to apply, in particular, to languages (such as the language of mathematical physics) that contain both *logical* rules such as the principles of arithmetic and analysis and *physical* rules such as Maxwell's field equations. For such languages the problem then is to distinguish precisely the two types of rules. Carnap proceeds in two steps. First, he defines a distinction between *logical* and *descriptive* expressions. The logical expressions, such as the connectives, quantifiers, and primitive signs of arithmetic, are those expressions such that all sentences built up from these expressions alone are determinate relative to the rules of the language.[12] For descriptive expressions such as the electromagnetic field functor, by contrast, whereas some sentences containing them are determined by the rules of the framework alone (e.g., Maxwell's equations themselves), this is not true for all such sentences (e.g., for sentences ascribing particular values of the electromagnetic field to particular space-time points). Intuitively, then, to determine the truth values of the latter sentences, we need extralinguistic information – such as, for example, observational information (§50). Given this distinction between logical and descriptive expressions, the distinction between logical and physical rules (and thus analytic and synthetic sentences) then follows easily (§51): the logical or analytic sentences are just those consequences of the rules of the formal language that contain only logical expressions essentially and thus remain consequences of the rules for all substitutions of nonlogical or descriptive vocabulary.[13]

11 According to Carnap's own explication of the analytic-synthetic distinction (see below), if a language is inconsistent, then there turn out to be no synthetic sentences (for all sentences are then determinate and all expressions logical). From Carnap's point of view, this is just one more respect in which an inconsistent language is an extremely inexpedient choice for the language of science, but it creates no fundamental difficulties for his underlying conception of analyticity. Compare Ricketts (1994, pp. 192–3).

12 A formal language is defined by formation rules and transformation rules. The transformation rules yield a consequence relation between sentences of the language. Sentences that are consequences of every sentence of the language are valid in that language. A sentence is determinate if it or its negation is valid.

13 It is this, for Carnap, that captures Wittgenstein's insight into the tautologousness of logical truth: the idea that logical truths hold in all conceivable circumstances and thus say nothing about the world. See Chapter 8 (this volume) and § V, this chapter. In *Introduction to*

In an earlier paper, I attempted to show that a problem closely related to the one raised by Gödel is fatal to this Carnapian explication of the concept of analyticity.[14] My point was that, if Carnap's explication is to have the desired result that classical arithmetic is analytic (relative to a suitable formal language), then the logical rules of the language in question have to include a non-recursively-enumerable (and indeed nonarithmetical) consequence relation; otherwise all arithmetical sentences are not, of course, determinate. But then, just as in Gödel's argument, we must have a meta-language that is essentially stronger than the object language in question, and Carnap's project again appears threatened by vicious circularity. One might view this argument, therefore, as an internal version of Gödel's argument. It attempts to show that Carnap's view of mathematics as conventional still founders on the incompleteness results, even when we work throughout with Carnap's explicitly language-relative version of the conventional/factual distinction.

Goldfarb and Ricketts (1992) also provide an extensive discussion of the issues raised by this latter argument. They question whether any vicious or otherwise objectionable circularity is involved in Carnap's use of a strong meta-language here. After all, Carnap himself is perfectly aware of the technical situation, and he explicitly states that the principle of tolerance is to be applied both at the level of the object language and at the level of the meta-language (§45), where, in particular, we are thus entirely free to use an "indefinite" (nonrecursive) notion of analyticity (compare §34a).[15] Goldfarb and Ricketts further object that my own attempt to explain why the Gödelian situation nonetheless presents a problem for Carnap proceeds by attributing to him a conception of logic that too closely assimilates his view

Semantics (1942, §16, 2a) this same idea is given a semantical reading: the logical or analytic truths remain true for all *interpretations* of the descriptive predicates. As Carnap explains, this change accommodates Tarski's work and, in particular, the existence of indefinable properties.

14 See Chapter 7 (this volume), which first appeared in 1988. I was not at the time acquainted with Gödel's drafts. I did, however, make essential use of the insightful paper by Beth (1963), which makes a point very similar to Gödel's.

15 Accordingly, Carnap (1935b/1953, p. 128) explicitly points out that "certain concepts referring to S_1 (e.g., 'analytic in S_1' ...) cannot be defined with the means of S_1 itself but only with those of a richer language S_2" – which assertion is proved in §60c of *Logical Syntax*. In (1935b/1953), Carnap cites Gödel's incompleteness paper and an earlier paper of his own, which reported the results of §60c before the appearance of the English translation. Indeed, already in §18 of *Logical Syntax*, Carnap qualifies his claim to have captured logical syntax within the object language: "[W]e shall *formulate the syntax of I* – so far as it is definite [recursive] – *in I itself*." In this sense, Carnap's anti-Wittgensteinian use of Gödel's method of arithmetization is also qualified (see footnotes 5 and 6, this chapter).

to the foundational conception of Hilbert. In particular, I suggested that Carnap's logicism requires that the meta-language, the language of logical syntax, should itself embody a purely combinatorial (and thus recursive) notion of analyticity or logical truth, so that Carnap in effect has two distinct notions of analyticity: a relativized, conventional notion for the various object languages and a privileged, combinatorial (and thus foundational in the sense of Hilbert) notion for logical syntax. Goldfarb and Ricketts argue, first, that there is no evidence at all for such a foundational conception in Carnap's text, and second, that the point of the principle of tolerance is precisely to wean us away from all such foundational concerns. According to Goldfarb and Ricketts, then, neither Gödel's original objection nor my internal variant takes adequate account of the absolutely central position of this principle in Carnap's philosophy of mathematics.[16]

II

Carnap first explicitly formulates the principle of tolerance in *Logical Syntax.* Yet, as he explains in his "Intellectual Autobiography," it actually represents a characteristic attitude toward philosophical problems that remained constant throughout his career:

> Since my student years, I have liked to talk with friends about general problems in science and practical life, and these discussions often led to philosophical questions. ... Only much later, when I was working on the *Logischer Aufbau,* did I become aware that in talks with my various friends I had used different philosophical languages, adapting myself to their ways of thinking and speaking. With one friend I might talk in a language that could be characterized as realistic or even materialistic; here we looked at the world as consisting of bodies, bodies as consisting of atoms .... In a talk with another friend, I might adapt myself to his idealistic kind of language. We would consider the question of how things are to be constituted on the basis of the given. With some I talked a language which might be labelled nominalistic, with others again Frege's language of abstract entities of various types, like properties, relations, propositions, etc., a language which some contemporary authors call Platonic.
>
> I was surprised to find that this variety in my ways of speaking appeared to some to be objectionable and even inconsistent. I had acquired insights

16 This is not to say, however, that Goldfarb and Ricketts see no serious problems arising for Carnap's philosophy of mathematics in the wake of Gödel's incompleteness results. On the contrary, they hold that Carnap's use of a strong meta-language to define analyticity has the damaging consequence of "not allow[ing] the conventional or non-factual nature of mathematics to be fully and explicitly displayed" (1992, p. 70), and they give qualified endorsement to the related criticism developed by Beth (footnote 14, this chapter).

> valuable for my own thinking from philosophers and scientists of a great variety of philosophical creeds. When asked which philosophical position I myself held, I was unable to answer. I could only say that my general way of thinking was closer to that of physicists and of those philosophers who are in contact with scientific work. Only gradually, in the course of the years, did I recognize clearly that my way of thinking was neutral with respect to the traditional controversies, e.g., realism vs. idealism, nominalism vs. Platonism (realism of universals), materialism vs. spiritualism, and so on.
>
> This neutral attitude toward the various philosophical forms of language, based on the principle that everyone is free to use the language most suited to his purpose, has remained the same throughout my life. It was formulated as "principle of tolerance" in *Logical Syntax* and I still hold it today, e.g., with respect to the contemporary controversy about a nominalist or Platonic language. (1963a, pp. 17–18; compare pp. 44–5)

Nevertheless, the particular ways in which Carnap attempted to implement this neutral or tolerant attitude varied with the problem situation in which he found himself.

In his very first publication, his doctoral dissertation of 1921–2, Carnap attempts to resolve the contemporary conflicts in the foundations of geometry involving mathematicians, philosophers, and physicists by carefully distinguishing among three distinct types or "meanings" of space: *formal*, *intuitive*, and *physical*. Carnap argues that the different parties involved in the various mathematical, philosophical, and physical disputes are, in fact, referring to different types of space, and, in this way, there is really no contradiction after all (1922, p. 64): "[all] parties were correct and could have easily been reconciled if clarity had prevailed concerning the three different meanings of space." Thus, mathematicians who maintain that geometry is purely logical or analytic are correct about formal space, philosophers who maintain that geometry is a synthetic a priori deliverance of pure intuition are correct about intuitive space, and physicists who maintain that geometry is an empirical science are correct about physical space.[17] In this way, in a *tour de force* of logical, mathematical, physical, and philosophical analysis, Carnap hopes to resolve the contemporary disputes about the foundations of geometry by showing how each of the conflicting parties – when they are limited to their proper domains – has a significant *part* of the truth.

As Carnap suggests above, however, it is in his next major publication, the *Aufbau* (1928a/1967), that his neutral attitude toward alternative "philosophical languages" comes fully into its own. And there are, in fact, two

17 As Carnap (1922, Chapter III) explains in detail, there is also a very substantial *conventional* element in our knowledge of physical space. For discussion, see Chapter 2, this volume.

importantly different aspects to this neutrality. In the first place, although Carnap develops one particular "constitutional system" in the *Aufbau*, the "system form with autopsychological basis," he also indicates the possibility of alternative systems, notably, the "system form with physical basis." Whereas the first logically reconstructs scientific knowledge from an epistemological point of view, by sketching a reduction of all scientific concepts to the given, the second logically reconstructs scientific knowledge from a materialistic or realistic point of view, by defining all scientific concepts (even those of introspective psychology) in terms of the fundamental concepts of physics. Both of these systems, according to Carnap, are equally possible and legitimate. In the second place, however, even within the domain of epistemology proper, Carnap also maintains an attitude of tolerance and neutrality toward the diverging, and apparently incompatible, philosophical epistemological schools:

> [*T*]*he so-called epistemological tendencies of realism, idealism, and phenomenalism agree within the domain of epistemology. Constitutional theory represents the neutral basis [neutrale Fundament] common to all. They first diverge in the domain of metaphysics and thus (if they are to be epistemological tendencies) only as the result of a transgression of their boundaries.* (1928a/1967, §178)[18]

Thus, since all epistemological schools agree that knowledge begins with the experiential given and then proceeds to build up all further objects and structures via a "logical progress," Carnap's autopsychological system (which does just this in a logically precise fashion) precisely represents what is clear and correct in all of them. The schools in question only disagree, therefore, when they indulge in metaphysical questions about which constituted structures are ultimately "real."

The vehicle of Carnap's philosophical neutrality, in the *Aufbau*, is the logic of *Principia Mathematica*. It is this system, understood in accordance with the logicist viewpoint he had first imbibed from Frege, that constitutes the fixed set of logical rules within which the various "philosophical languages" he considers are then formulated. When he became involved in the disputes on the foundations of logic and mathematics in the late 1920s, however, Carnap could no longer persist in this state of happy logical innocence.[19]

18 For further discussion of both aspects of Carnap's neutrality in the *Aufbau*, see Part Two (this volume).

19 Although the *Aufbau* was first published in 1928, most of the work on it was completed in the years 1922–5, before Carnap moved to Vienna to join Schlick's Circle in 1926. It was only after moving to Vienna that Carnap became involved in the disputes on the foundations of mathematics. See (1963a, pp. 16–20, 46–50).

For he was now faced with a situation in which the background rules of logic were precisely what was at issue. Even worse, logic and mathematics were now embroiled with philosophical questions about mathematical intuition, the "reality" of mathematical objects, and the relation of such objects to the thinking subject–just the kind of questions that logicism had hoped to be done with once and for all. Fruitless and interminable philosophical disputes, which Carnap himself had hoped to avoid through the tolerance and neutrality of the *Aufbau* project, were now threatening the very basis of that project. But how can disputes about the foundations of logic themselves be logically resolved?

Carnap's first idea was to incorporate the apparently conflicting demands of intuitionism and formalism within logicism. Carnap sketches this idea in contributions to two symposia on the foundations of mathematics in 1930.[20] With respect to intuitionism, we drop the purely philosophical doctrine that "arithmetic rests on an original intuition [*Ur-Intuition*]" while retaining only the "*finitist-constructivist* requirement ... [of] renouncing pure existence proofs without constructive procedures" – we thereby obtain an accommodation with formalism, which also recognizes this constraint in the realm of metamathematics (1930b, pp. 308–9). Logicism, by contrast, has run into problems concerning the need for special existence assumptions expressed in the axioms of infinity, choice, and reducibility. Carnap is happy to adopt Russell's expedient of considering infinity and choice as nonlogical premises or conditions in theorems for which they are needed, and he here focuses his efforts at reconciliation on the axiom of reducibility. Although Ramsey has made an excellent case for rejecting the ramified theory of types in favor of the simple theory, his justification for impredicative definitions embodies an "absolutistic" and "theological" assumption of the existence, prior to any definition or construction, of the totality of all properties – which assumption, however, is clearly incompatible with the "finitist-constructivist requirement" (*ibid.*).[21] Carnap's own not fully developed countersuggestion is to restrict ourselves to finitely definable properties while still retaining at

20 The first was prepared for a symposium on the philosophical foundations of mathematics in the *Blätter für deutsche Philosophie* 4 (1930), wherein intuitionism was treated by Menger (footnote 3, this chapter) and formalism by Paul Bernays. Carnap's contribution is "Die Mathematik als Zweig der Logik"(1930b). The second (1931a) is the better-known paper, "Die logizistische Grundlegung der Mathematik," presented to the Second Conference on the Epistemology of the Exact Sciences in Königsberg in September 1930. This paper appears in the same issue of *Erknenntnis* as papers by the co-symposiasts Arend Heyting representing intuitionism and John von Neumann representing formalism. All three of the latter papers are translated in Benacerraf and Putnam (1983).

21 See also Carnap (1931a/1983, p. 50).

least the most important impredicative definitions (1931a/1984, pp. 50–2). Finally, the chief remaining difference between logicism and formalism is that logicism develops definitions of the natural numbers via properties or classes, whereas formalism considers the numerals as primitive signs. If, however, we reflect on the need to account also for the application of arithmetic, and thus to construct a formal system in which empirical statements involving numbers are also derivable, then, Carnap suggests, a formal system meeting this desideratum might very well lead us back to the logicist definitions (1930b, pp. 309–10).[22] In this way, Carnap hopes, we may attain "a problem-solution that will appear as satisfactory from [all three] different points of view" (1930b, p. 310).[23]

Yet Carnap's idea of articulating a single formal-logical system that would thus simultaneously fulfill the demands of all three foundational schools was never successfully carried out.[24] Instead, he adopts the fundamentally new standpoint encapsulated in the principle of tolerance in *Logical Syntax.* The way to dissolve the fruitless foundational disputes is not to develop a single logical system simultaneously embracing the demands of all parties. Rather, we should view the choice of underlying logic, too, as simply the choice of one form of language among an infinity of equally possible alternatives. Intuitionism is correct that we can, if we wish, develop a language, Language I, embodying finitist-constructivist restrictions on existential quantification. But logicism is equally correct that we can also, if we wish, develop a much stronger language, Language II, in which the full unrestricted existence claims of classical analysis and set theory are analytic. Indeed, by employing the device of "coordinate languages" in which numerical expressions

22 See also (1931b, pp. 141–4) for Carnap's more detailed presentation of this idea.

23 See also Carnap's (1931b, p. 141) remarks introducing his discussion: "Many listeners have received the depressing impression from the three lectures that the problem-situation is tangled, confused, and hopeless: here are three tendencies, none of which understand any of the others and each of which wants to construct mathematics in a different way. But in reality the situation is not as bad as this, as we will see." He concludes (p. 144): "I believe that this execution [of the ideas of all three schools] will finally lead to a common result."

24 Carnap does not explain why he gave up on this idea, but we may plausibly conjecture that interaction with Tarski and Gödel convinced him that the definability restrictions on arithmetical properties he had envisioned would not lead to a satisfactory version of classical analysis. He may have been influenced, in this regard, by Tarski's work during 1930–1 on definable sets of real numbers, which he cites in the bibliography to the English version of *Logical Syntax.* It is not clear what the relation is between the finitist-constructivist ideas of Carnap's 1930 contributions and the early versions of *Logical Syntax* restricted to Language I alone (footnote 3, this chapter). Carnap (1963a, p. 33) notes that he had planned to develop a version of the "Zermelo-Fraenkel axiom system of set theory, but restricted in the sense of a constructivist method" already in 1927.

appear as the basic individual constants, we can, since such numerical expressions are logical in the sense of §50, even count the axiom of infinity as logical (§38a). Nor is there any need to scruple over the admissibility of impredicative definitions, for there is again only the question of what form of language we *wish* to adopt.[25] Further, and for kindred reasons, the axiom of choice is also a perfectly admissible, although optional, logical principle (§§34h, 38a). Finally, we can easily reconcile the demands of logicism and formalism – not, however, by an argument that a single formal system embracing the application of arithmetic must eventually lead us back to the logicist definitions of the arithmetical terms, but simply by the mere possibility of a single "*total language* that unites the logico-mathematical and the synthetic sentences" (§84).[26]

III

As we noted at the beginning (footnote 3, this chapter), Carnap credits Karl Menger's 1930 paper, "Der Intuitionismus," for first representing the standpoint of tolerance within the foundations of mathematics:

> One may assume that the tolerant attitude intended here, applied to special mathematical calculi, comes naturally to most mathematicians without customarily explicitly articulating it. In the conflict over the logical foundations of mathematics it has been represented with particular force (and apparently for the first time) by *Menger* [*Intuitionismus*], pp. 324f. *Menger* points out that the concept of constructivity, which intuitionism absolutizes, can be taken narrowly or widely. – How important it is for the clarification of philosophical pseudo-problems also to apply the attitude of tolerance to the form of the total language will become clear later (cf. §78). (§17)

25 See §44: "One can permit such definitions or exclude them, without giving a justification. But if one wants to justify the one or the other procedure, then one must first exhibit the formal consequences of this procedure." Goldfarb and Ricketts (1992, p. 68; compare pp. 62–3) note the contrast between this attitude and that of "Die logizistische Grundlegung der Mathematik."

26 Indeed, as noted at the very beginning, Carnap does not pursue the traditional logicist project of defining the arithmetical terms via logical terms in the earlier sense (connectives, quantifiers, identity) at all in *Logical Syntax.* He instead treats the arithmetical terms as primitive in both Language I and Language II. They nonetheless count as logical in the new sense of §50. Carnap observes in §84 that the question of logicism in the traditional sense is not even well defined: "[W]e have given a formal distinction between logical and descriptive signs in general syntax; but a sharp division of the logical signs in our sense into logical signs in the narrower [traditional] sense and mathematical signs has not yet been given by anyone." Goldfarb and Ricketts (1992, p. 68) note the resulting attenuation of traditional logicism.

On reflection, however, this passage helps us rather to articulate what is entirely unique in Carnap's own understanding of the principle of tolerance.

Menger's representation of "tolerance" appears in §10 of his paper, entitled "General Epistemological Remarks" (1930/1979, pp. 56–8). Referring to his own earlier work on the set-theoretical meaning of various ideas of Brouwer's, Menger suggests that one might develop constructivity requirements corresponding to the admissibility of stronger and stronger sets – finite sets, denumerable sets, analytic sets, and finally arbitrary sets of real numbers. In this way, one can envision a variety of systems meeting a variety of constructivity requirements (the weakest being mere consistency). There is then no need, as intuitionists customarily do, to attach oneself dogmatically to one particular notion of constructivity (p. 57): "[f]or in mathematics and logic it does not matter which axioms and principles of inference one *assumes*, but rather what one can *derive* from them or with their help respectively." According to this "implicationist" standpoint, then, we are concerned only with the purely mathematical problem of which consequences follow from which given assumptions. We are interested in the "mathematics" of constructivity, not in the purely "biographical" question of which principles appeal to which actual mathematicians.

Menger's attitude toward the philosophical debate over the foundations of logic and mathematics is therefore one of stark dismissal: let us put all such merely "dogmatic" and "biographical" questions aside once and for all and simply get on with the real mathematical work. In this sense, Menger perfectly represents the attitude of the "ordinary working mathematician" to which Carnap alludes above. Yet Carnap's own attitude is very different, for he, unlike Menger, is intensely interested in the philosophical foundational debate. Like Menger, to be sure, he wants to do away once and for all with "the dogmatic attitude through which the discussion often becomes unfruitful" (§16). But the whole point of Carnap's principle of tolerance is to articulate a systematic method for resolving or dissolving such philosophical disputes. Carnap's principle, we might say, is crafted for and directed at philosophers: it aims to offer (scientifically minded) philosophers a way out of their impasses and perplexities. From a purely mathematical point of view, of course, Carnap's constructions are of very limited interest. In this sense, Carnap's principle is not a call, like Menger's, for mathematicians to leave behind philosophy.[27]

27 In his introduction to the translation, Menger (1979, pp. 13–14) makes several revealing remarks in his anxiety to establish exclusive priority for the idea of tolerance. Thus, he first addresses Carnap's attribution, in the above passage from §17 of *Logical Syntax*,

Carnap's reference to §78 signals precisely this difference between his own attitude and Menger's, for here Carnap puts forward nothing less than a general characterization of the peculiar type of confusion arising in philosophy:

> That in philosophical debates, even in those that are free of metaphysics, unclarities occur so frequently, and that in philosophical discussions there is so much talk at cross purposes, is due for the most part to the use of the material mode of speech instead of the formal [mode]. (§78)[28]

We are misled, in the first place, to think that we are debating about "extra-linguistic objects, such as numbers, things, properties, experiences, states of affairs, space, time, etc." instead of about "language structures and their interconnections . . . such as numerical expressions, thing-designations, spatial coordinates, etc." We are thereby misled, in the second place, to ignore the *relativity to language* that is a central feature of the formal or syntactic concepts. We thus ignore the all-important circumstance that properly formal or syntactic claims must first specify the language in question: they can apply to all languages, some languages, one given language, or (perhaps most interestingly) they can serve as *proposals* to formulate the total language of science (or some part thereof) in one or another particular fashion.

Carnap immediately applies these ideas to a debate in the foundations of mathematics, namely, the debate between logicism and formalism. If the logicist asserts, in the material mode, that "numbers are classes of classes of things," and the formalist asserts, also in the material mode, that "numbers belong to a peculiar, original type of objects," we are hopelessly stuck:

> Then between the two an endless and fruitless discussion can be carried out over who is correct and what the numbers really are. The unclarity vanishes if the formal mode of speech is applied.

of an attitude of tolerance to "most mathematicians." To this Menger remarks that the "prominent mathematicians . . . who have dealt with the foundations of mathematics," such as Poincaré, Hilbert, Weyl, and Brouwer, have been quite opposed to it – thereby missing Carnap's point, namely, that the "ordinary working mathematician" tacitly embraces tolerance. Menger then considers the passage from Carnap's intellectual autobiography, cited at the beginning of §II, this chapter, in which Carnap says that the attitude of tolerance "has remained the same throughout my life." Here Menger simply remarks that Carnap's memory must have been faulty. In this way, Menger not only misses the point that it is only specifically *logico-mathematical* tolerance that is new in *Logical Syntax* but also the deep roots of Carnap's principle in a much more general attitude toward philosophical problems as such.

28 Compare §75: "[By means of the diagnosis of the material mode of speech] the character of philosophical problems in general will become clear. The unclarity about this character is traceable mainly to the deception and self-deception brought about via the application of the material mode of speech."

Accordingly, we translate the above two sentences into "the numerical expressions are class expressions of the second level" and "the numerical expressions are expressions of the zeroth level," respectively. It now becomes clear that we are talking about two different languages, and, since both languages are perfectly possible, "the dispute vanishes." The only remaining possibility, therefore, is that

> the discussants understand one another as intending their theses as suggestions [for the language of science]. In this case one cannot debate about the truth and falsehood of the theses, but only whether this or that form of speech is simpler or more suitable for such and such purposes.

In this way, we offer our (scientifically minded, nonmetaphysical) philosophical friends a way of transforming their fruitless dispute into a fruitful one. They are not really debating about the "true nature" of mathematical objects but merely proposing different language forms, each having various advantages and disadvantages, for the total language of science.

It cannot be stressed too much, I think, that this diagnosis and transformation of characteristically philosophical problems constitutes the main point of both the principle of tolerance and the method of logical syntax more generally. Thus, in his "Intellectual Autobiography," Carnap states that "the investigation of philosophical problems was originally the main reason for the development of syntax," and he expands on this as follows:

> [I]t seemed to me important to show that many philosophical controversies actually concern the question whether a particular language form should be used, say, for the language of mathematics or of science. For example, in the controversy about the foundations of mathematics, the conception of intuitionism may be construed as a proposal to restrict the means of expression and the means of deduction of the language of mathematics in a certain way, while the classical conception leaves the language unrestricted. I intended to make available in syntax the conceptual means for an exact formulation of controversies of this kind. Furthermore, I wished to show that everyone is free to choose the rules of his language and thereby his logic in any way he wishes. This I called the "principle of tolerance"; it might perhaps be called more exactly the "principle of the conventionality of language forms". As a consequence, the discussion of controversies of the kind mentioned need only concern, first, the syntactical properties of the various forms of language, and second, practical reasons for preferring one or the other form for given purposes. In this way, assertions that a particular language is the correct language or represents the correct logic such as often occurred in earlier discussions, are eliminated, and traditional ontological problems, in contradistinction to the logical or syntactical ones, for example, problems about "the essence of number", are entirely abolished. (1963a, pp. 54–5)

Carnap is perfectly serious about this: traditional philosophy should be replaced by the new and logically exact enterprise of "language planning" (1963a, §11, especially pp. 67–9). Only so can we achieve an exact diagnosis of the true character of traditional philosophical problems and, at the same time, find a new (albeit still characteristic) task for the philosophy of the future.[29]

Carnap thus adopts a deflationary stance toward traditional philosophy, but it is nonetheless a characteristically philosophical form of deflationism. Carnap does not simply leave philosophy behind in favor of the standpoint of the "working scientist." Rather, he systematically articulates a radically new vision of the philosophical enterprise, in which, in particular, philosophy is to retain its special, nonempirical status:

> Metaphysical philosophy claims to go beyond the empirical-scientific questions of a scientific domain and pose questions about the essence of the objects of the domain. The non-metaphysical logic of science also takes up a different standpoint than that of empirical science; not, however, by means of a metaphysical transcendence, but rather by the circumstance that it makes the linguistic forms themselves the objects of a new investigation. (§86)

In this way, we obtain a radically new conception of philosophical problems and, in particular, of the true character of the philosophical debate in the foundations of logic and mathematics. It is this transformation and reformulation of the philosophical debate with which Carnap (in sharp contrast to a "working scientist" such as Menger) is most concerned.[30]

29 Ricketts (1994, and especially 1982), correctly emphasizes Carnap's concern with the difference between genuine rational disputes and traditional philosophical talk at cross purposes. But he then characterizes Carnap's main problem as that of applying this distinction to actually occurring intellectual debates so as to determine which type of case we are faced with in fact: there is a genuine dispute if and only if the investigators in question share a common language or linguistic framework. From this point of view, Carnap immediately runs up against a difficult problem in *descriptive* syntax, namely, how to tell whether or not actual investigators share a common language, and Quine's challenge – that, in Ricketts's terms, of supplying a "criterion of analyticity" – then proves to be fatal to the entire enterprise. I do not think this fits Carnap's own conception of his project. Carnap is not worried about determining, in actual cases, which disputes are genuine and which are not. He is already perfectly clear about this: *philosophical* disputes are characteristically fruitless, whereas *scientific* questions (in either natural science or mathematics) patently are rationally negotiable. Carnap's problem is not to discriminate the fruitless disputes from the fruitful ones but to offer those enmeshed in the former – philosophers – a way out. This is what the construction and investigation of a variety of formal languages is for. Descriptive syntax can fall where it may: Carnap is concerned with the constructive task – belonging to *pure* syntax – of language planning.

30 Carnap tirelessly reiterates this new conception of the true character of philosophical problems, normally with the debate in the foundations of mathematics as paradigmatic. See,

IV

This transformation and reformulation of traditional philosophy involves Carnap himself in a philosophical task. How do we precisely characterize the distinction between questions that do concern the "true natures" of objects (questions investigated in natural science and mathematics) and those that merely concern forms of language (questions for philosophy)? How do we show our (scientifically minded, nonmetaphysical) philosophical friends that their problems are actually of the second kind? In §§76–7 of *Logical Syntax* the distinction is drawn with the help of the concept of *universal words* [*Allwörter*]. Formally or syntactically considered, a universal word is a predicate of a language such that every predication thereof is logically or analytically true in that language. Thus "number" is a universal word in the language of arithmetic whereas "prime number" is not, "being a space-time point" is a universal word in the language of mathematical physics (§40) whereas "being a space-time point characterized by such-and-such value of mass-density" (or charge-density, or electromagnetic field, etc.) is not. From a formal or syntactic point of view, universal words are entirely dispensable, for they can always be replaced by distinctive types of variables. In philosophical discussions, however, we characteristically find universal words used in the material mode of speech:

> The investigation of universal words is especially important for the analysis of philosophical sentences. They occur very frequently in such sentences, both in metaphysics and in the logic of science, and mostly in the material mode of speech. (§76)

It is this that misleads us into asking questions about the "reality" or "true nature" of numbers, the "reality" or "true nature" of space-time points. The syntactic transformation in the formal mode, by contrast, makes it clear that we are really posing questions about the form of language and, in particular, about what types of variables are to occur at various levels.[31]

Here we have the germ of Carnap's (1950a/1956) celebrated later distinction between *internal questions*, which are rationally answerable on the basis of the rules of a given language or linguistic framework, and *external questions*, which rather concern the prior choice of one or another such framework as the language for the investigation in question. External questions are therefore noncognitive or nontheoretical, and concern only the

e.g., Carnap (1934a; 1934b, especially §7 of the Supplementary Remarks; 1935a, especially pp. 75–82).

31 In §76, Carnap traces this idea to Wittgenstein's doctrine of "formal concepts" in the *Tractatus*.

purely practical problem of which framework is adapted or expedient for one or another given purpose. Questions of "reality" or existence thus make theoretical sense within a given framework as internal questions ("Is there a prime number greater than a hundred?" "Are there space-time points having such-and-such values of mass-density?"), but they have no such sense taken as external questions ("Are there really numbers?" "Are there really space-time points?").[32] Carnap applies these notions, once again, to the philosophical problem of the foundations of mathematics:

> [P]hilosophers who treat the question of the existence of numbers as a serious philosophical problem ... might try to explain what they mean by saying that it is a question of the ontological status of numbers; the question whether or not numbers have a certain metaphysical characteristic called reality (but a kind of ideal reality, different from the material reality of the thing world) or subsistence or status of "independent entities". Unfortunately, these philosophers have so far not given a formulation in terms of the common scientific language. Therefore our judgement must be that they have not succeeded in giving to the external question and to the possible answers any cognitive content. Unless and until they supply a clear cognitive interpretation, we are justified in our suspicion that their question is a pseudo-question, that is, one disguised in the form of a theoretical question while in fact it is non-theoretical; in the present case it is the practical problem whether or not to incorporate into the language the new linguistic forms which constitute the framework of numbers. (1950a/1956, §2, p. 209)

The kinship with the program of *Logical Syntax* is therefore evident.[33]

In the philosophical debate in the foundations of mathematics, questions about the existence of numbers (and higher set-theoretical objects) arose in connection with the need for strong existential axioms (such as infinity, choice, and reducibility) in the wake of the discovery of the paradoxes. Carnap's remarks on "Existence Assumptions in Logic" in §38a of *Logical Syntax* are therefore of particular interest. In the case of the axiom of infinity, for example, which is demonstrable in both Language I and Language II, Carnap holds that we are here concerned only with the choice of a so-called coordinate language, in which numerical expressions are of zeroth type. In such languages, numerical expressions are logical rather

32 Carnap (1950a/1956, §2, pp. 209, 213) points out that "There are numbers" and "There are space-time points" also have entirely trivial internal readings on which they are obviously analytically true, but these are not questions in which anyone – philosopher or scientist – is seriously interested.

33 Carnap (1950a/1956, §3, pp. 213–14) explains that external questions primarily concern the choice of a distinctive type of variable.

than descriptive in the sense of §50, and the axiom of infinity therefore counts as an analytic truth. For precisely this reason, there can be no genuine ontological issue here:

> The [sentences containing only logical expressions] (and with them all sentences of mathematics) are, from the point of view of material interpretation, expedients for the purpose of operating with the [sentences containing descriptive expressions]. Thus, in laying down [a sentence containing only logical expressions] as a primitive sentence, only usefulness for this purpose is to be taken into consideration. (§38a)[34]

There can be no genuine theoretical question whether a primitive mathematical existence assertion is acceptable or not *precisely because* such sentences contain only logical expressions. If we add such a sentence to our language, we obtain merely a new analytic truth, and whether a language with this or that primitive analytic truth is acceptable or not can only be a purely pragmatic question.[35] In this sense, "existence assumptions in logic" must, in Carnap's later terminology, count as external questions.

The situation is quite otherwise in the empirical sciences. In the language of mathematical physics (§82), for example, we postulate both logical rules (L-rules) and physical rules (P-rules), where the latter consist customarily of "certain most general laws" called "*fundamental laws* [*Grundgesetze*]" and their logical consequences (p. 316[244]). Maxwell's equations for the electromagnetic field are paradigmatic of such "P-fundamental sentences" (p. 319[247]). Just as in the case of "existence assumptions in logic," then, there is a question whether or not to add such primitive sentences to the rules of our language. Here, however, we are *not* faced with a purely pragmatic, external question. For such a P-fundamental sentence, like any other sentence containing descriptive expressions essentially, can and must be empirically tested:

> A sentence of physics, whether it is a P-fundamental sentence or an otherwise valid sentence or an indeterminate assumption (i.e., a premise whose

34 As Carnap points out, this applies also to the existential axioms of set theory. Similarly, in the language of mathematical physics, sentences asserting the existence of (a nondenumerable infinity of) space-time points are similarly analytic and thus devoid of genuine ontological import. As we noted at the end of §II, this chapter, this attitude toward mathematical existential assumptions marks a sharp break from Carnap's earlier treatment.

35 As Goldfarb has emphasized to me in discussion, the concept of "primitive sentence" is not well defined on the basis of a notion of *consequence* (or validity) alone. In this connection, especially, Carnap himself also employs a notion of *derivability* in characterizing formal languages (§§13, 14, 30–3, 47–8). In any case, the heart of the above point can be formulated without relying on the notion of primitive sentence: whether a language in which this or that analytic sentence is valid is acceptable or not can only be a purely pragmatic question.

> consequences are investigated), is *tested*, in that consequences are deduced from it on the basis of the transformation rules of the language until one finally arrives at propositions of the form of protocol-sentences. These are compared with the protocol-sentences actually accepted and either confirmed or disconfirmed by then. If a sentence that is an L-consequence of certain P-fundamental sentences contradicts a proposition accepted as a protocol-sentence, then some alteration must be undertaken in the system. (p. 317[245])

To be sure, what precise change we then make is not itself determined by rules, and faced with such a situation we might even make a change in the L-rules. Nevertheless, there remains an essential distinction:

> If we assume that a newly appearing protocol-sentence within the language is always synthetic, then there is nonetheless the following difference between an L-valid and thus analytic sentence S_1 and a P-valid sentence S_2, namely, that such a new protocol-sentence – whether or not it is acknowledged as valid – can be at most L-incompatible with S_2 but never with S_1. (pp. 318–19[246])[36]

In this sense, "[t]he laws have the character of *hypotheses* relative to the protocol-sentences" (p. 318[245]), and such hypotheses, despite their postulational character as primitive rules of the language of mathematical physics, "are to be tested by empirical material, i.e., by the actually present and ever newly added protocol-sentences" (p. 320[248]). Therefore, the conventional element in the adoption of P-fundamental sentences is strictly limited. It does not derive, as in the case of L-rules, from the utter logical irrelevance of empirical material, but rather from the circumstance of empirical underdetermination.[37]

It follows that the question of adopting a given P-fundamental sentence – despite the fact that such a sentence, like the fundamental logico-mathematical sentences, is definitive of the rules of the language – is not, in Carnap's later terminology, a purely external question. The answers to internal questions, Carnap says (1950a/1956, §2, pp. 206–7), "may be found either by purely logical methods or by empirical methods," and, in the latter case, "[r]esults of observation are evaluated according to certain rules as

36 The translation inadvertently has "incompatible" rather than "L-incompatible" here.

37 See p. 320[249]: "That there is still always a conventional element in the hypotheses, despite their subjection to empirical control by the protocol-sentences, rests on the circumstance that the system of hypotheses is never uniquely determined by the empirical material, no matter how rich." Ricketts (1994, pp. 193–5) interprets Carnap's empiricism as the requirement that *indeterminate* sentences be testable via protocol sentences, whereas Carnap himself clearly extends this requirement to all *synthetic* sentences, including, in particular, the P-rules.

confirming or disconfirming evidence for possible answers." Such "rules of evaluation" may, as in *Logical Syntax*, consist merely in the hypothetico-deductive method. In this case, the relevant rules are clearly analytic. Or, as in *Logical Foundations of Probability* (Carnap 1950b), we might incorporate a confirmation function into our language. Here, again, however, our "rules of evaluation" are still analytic.[38] In all cases, then, the rules definitive of internal questions are logical or analytic rules: it is precisely the possibility of coming to a decision on the basis of such rules that make a question more than purely pragmatic.[39] Hence, what is crucial, for Carnap, is not the bare idea of a formal language or linguistic framework as such. After all, any scientific decision whatsoever, even whether or not to accept a given empirical theory, can be represented as the choice of a particular formal language. What is crucial is the distinction, *within* any formal language or linguistic framework, between analytic and synthetic sentences. It is because analytic sentences (and therefore L-rules) are true solely in virtue of meaning whereas synthetic sentences (and therefore P-rules) must also respect

38 See Carnap (1950b/1962), p.v): "all principles and theorems of inductive logic are analytic." As Goldfarb and Ricketts have emphasized to me in discussion, there is another possible reading of the second passage from "Empiricism, Semantics, and Ontology" just cited, according to which the "rules of evaluation" in question include P-rules (such as general laws of nature) as well as L-rules. In view of Carnap's explicit use of the notion of *confirmation* here, together with the circumstance that "Empiricism, Semantics, and Ontology" belongs to the same period as *Logical Foundations of Probability*, I consider this reading to be less plausible (so that, on my reading, the examples of internal questions in the "thing language" that Carnap presents here should be conceived of as singular predictions not requiring laws of nature). In any case, however, what is central to my argument is that Carnap nowhere (to my knowledge) suggests a confirmational asymmetry between P-rules and indeterminate synthetic sentences, whereas he consistently maintains a clear confirmational asymmetry between P-rules and analytic sentences (see footnote 39, this chapter).

39 I am therefore entirely in agreement with Ricketts's (1982, p. 123) description of the situation: "Observation reports can confirm and disconfirm theories within frameworks, but they can never confirm or disconfirm the logical machinery of a framework. This logical machinery is required to constitute a framework-relative notion of evidence. Only against the background of such a notion does talk of confirmation make any sense at all. So, in changes of theory, the application of pragmatic considerations is confined to confirmationally acceptable theories and thus governed by the logical machinery of the framework. In changes of framework, however, pragmatic considerations operate untrammeled: there is nothing else." What I am adding here is simply the observation that P-rules are subject to the same rules of confirmation as are indeterminate synthetic sentences. In more recent work, by contrast, Ricketts uses the point that P-rules are just as definitive of a formal language as are L-rules to urge that we should divorce the distinction between change of theory and change of language (and the related distinction between internal and external questions) from the analytic/synthetic distinction: see, e.g., (1994, p. 189).

the empirical facts that changes in the former, but not the latter, are purely pragmatic.[40]

V

Following out the implications of Carnap's own understanding of the principle of tolerance has therefore led us back to the absolutely central position of the analytic/synthetic distinction in his philosophy. Carnap's tolerance is not simply that of the "working scientist," who urges us to leave philosophical problems behind once and for all in order to return to the real scientific work. Rather, it is directed precisely at those caught in serious philosophical perplexities, and it aims to offer such people (provided, of course, that they are inclined toward scientific rather than metaphysical philosophizing) a way of transforming their hitherto fruitless disputes into fruitful ones. We are invited, in particular, to recognize the true character of philosophical problems as questions about the logico-linguistic form in which the total language of science is to be cast. They are not genuine theoretical questions, such as are treated in the mathematical and natural sciences themselves, but purely pragmatic external questions governed by canons of expedience rather than truth. And, as we have seen, what shows us that such external questions really are purely pragmatic is precisely the circumstance that they concern, in the end, only the question of which primitive *analytic* sentences to adopt. It is for precisely this reason, that is, that such questions involve us with no "matters of fact."

Thus, in the case of philosophical problems in the foundations of mathematics, their true character is revealed when we recognize mathematical sentences as mere formal auxiliaries within the total language of science:

> The application of synthetic and analytic sentences in science is as follows. Factual science lays down synthetic sentences, e.g., singular sentences for the description of observed facts or general sentences that are laid down as hypotheses and are applied experimentally. From the sentences thus laid

40 See Carnap (1963b, p. 921). Here, Carnap considers Quine's argument that no statement of science is immune from revision. Carnap admits (as §82 of *Logical Syntax* had already stated explicitly) that logico-mathematical rules – just like physical rules – can be revised, and in both cases one has "a transition from a language L_n to a new language L_{n+1}." However: "My concept of analyticity as an explicandum has nothing to do with such a transition. It refers in each case to just one language .... That a certain sentence S is analytic in L_n means only something about the status of S within the language L_n; as has often been said, it means that the truth of S in L_n is based on the meanings in L_n of the terms occurring in S."

> down the scientist now tries to derive other synthetic sentences, e.g., to make predictions about the future. The analytic sentences serve as auxiliaries for these inferential operations. Considered from the point of view of the total language, the whole of logic, including mathematics, is nothing else but an auxiliary calculus for handling synthetic sentences. *Formal science* has no independent meaning. It is rather introduced into the language as an auxiliary component on technical grounds, so as to make the linguistic transformations required for *factual science* technically easier. The great importance pertaining to formal science, and thus to logic and mathematics, in the total system of science is thereby in no way denied but rather precisely emphasized, through a characterization of the particular function [of this science]. (Carnap 1935b/1953, p. 127)[41]

And it is precisely in virtue of this "particular function" of logic and mathematics as mere deductive auxiliaries that we then can apply the principle of tolerance here:

> [I]f we regard interpreted mathematics as an instrument of deduction within the field of empirical knowledge rather than as a system of information, then many of the controversial problems are recognized as being questions not of truth but of technical expedience. The question is: Which form of the mathematical system is technically most suitable for the purpose mentioned? Which one provides the greatest safety? If we compare, e.g., the systems of classical mathematics and of intuitionistic mathematics, we find that the first is much simpler and technically more efficient, while the second is more safe from surprising occurrences, e.g., contradictions. (Carnap 1939, §20)

Hence, without clear and precise distinctions, within the total language of science, between logical and descriptive expressions, logical and physical rules, analytic and synthetic sentences, we could not use the principle of tolerance to dissolve the philosophical disputes in question.[42] So it is no

41 Compare also Carnap (1939, §§1, 7, 23).

42 This explicit link between the principle of tolerance and the analytic/synthetic distinction marks the central difference between the present interpretation and the viewpoint of Goldfarb and Ricketts. I believe that this difference is traceable, in the end, to the circumstance that Goldfarb and Ricketts are operating against the background of Ricketts' general understanding of the principle of tolerance, according to which Carnap's central concern is that of determining, in the context of *descriptive* syntax, whether or not two different investigators share a common linguistic framework. From this point of view – the point of view of Quine's demand for a "criterion of analyticity" (cf. Goldfarb and Ricketts [1992, p. 75, n. 21]) – the real problem concerns the general notion of linguistic framework as such, together with the accompanying notion of "true in virtue of the adoption of a framework." From this point of view, therefore, the distinctions, within a given framework, and in *pure* syntax, between logical and descriptive expressions, logical and physical rules, are of decidedly secondary importance. See footnotes 29, 37, and 39, this chapter.

wonder, then, that Carnap continually reiterates the importance of these distinctions.[43]

We are thereby led back to Carnap's logicism and, in particular, to his debt to Frege:

> [T]he following conception, which derives essentially from Frege, seemed to me of paramount importance: It is the task of logic and mathematics within the total system of knowledge to supply the forms of concepts, statements, and inferences, forms which are then applicable everywhere, hence also to non-logical knowledge. It follows from these considerations that the nature of logic and mathematics can be clearly understood only if close attention is given to their application in non-logical fields, especially in empirical science. Although the greater part of my work belongs to the fields of pure logic and the logical foundations of mathematics, nevertheless great weight is given in my thinking to the application of logic to non-logical knowledge. This point of view is an important factor in the motivation of some of my philosophical positions, for example, for the choice of forms of languages, for my emphasis on the fundamental distinction between logical and non-logical knowledge. (1963a, pp. 12–13)

And, as Carnap explains (p. 12), this Fregean view that "knowledge in mathematics is analytic in the general sense that it has essentially the same nature as knowledge in logic" later "became more radical and precise, chiefly through the influence of Wittgenstein." For it was Wittgenstein, according to Carnap, who first taught him that logic (and therefore mathematics) is entirely independent of all "matters of fact":

> The most important insight I gained from [Wittgenstein's] work was the conception that the truth of logical statements is based only on their logical structure and on the meaning of the terms. Logical statements are true under all conceivable circumstances; thus their truth is independent of the contingent facts of the world. On the other hand, it follows that these statements do not say anything about the world and thus have no factual content. (p. 25)

43 Thus, for example, Carnap (1963b, p. 932) considers Tarski's view that there is no sharp distinction between logical and descriptive expressions: "[This disagreement] is to a large extent to be explained by the fact that Tarski deals chiefly with languages for logic and mathematics, thus languages without descriptive constants, while I regard it as an essential task for semantics to develop a method applicable to languages of empirical science. I believe that a semantics for languages of this kind must give an explication for the distinction between logical and descriptive signs and that between formal and factual truth, because it seems to me that without these distinctions a satisfactory methodological analysis of science is not possible." And, more simply, Carnap (1942) states, in §13, which is concerned precisely with the distinction between logical and descriptive signs, that "[t]he problem of the nature of logical deduction and logical truth is one of the most important problems in the foundations of logic and perhaps in the whole of theoretical philosophy."

Wittgenstein's doctrine of tautology is thus the fulfillment, for Carnap, of Frege's logicism (compare p. 46). And logicism so understood is therefore an integral part of Carnap's own understanding of the principle of tolerance.[44]

Wittgenstein's doctrine of tautology rests on a sharp distinction between the logical constants and all other meaningful signs. Tautologies remain true for all combinations of existence and nonexistence of states of affairs. However, that the logical signs themselves are held constant in this process of evaluation does not limit the resulting independence of logic from the totality of facts constituting the world. For the logical constants, unlike all other primitive signs, are not representative of objects:

> The possibility of the proposition rests on the principle of the representation of objects by means of signs.
>
> My fundamental thought is that the "logical constants" are not representative. That the *logic* of the facts can not be represented. (4.0312)

The logical constants obtain this uniquely privileged status in virtue of the circumstance that all meaningful propositions, for Wittgenstein, are the results of truth operations on elementary propositions. All meaningful propositions arise by iteratively applying the operations of truth-functional composition and quantification to a given initial collection of propositions that are themselves logically simple and thus contain no logical constants. In this sense, the logical constants "vanish," since they merely afford us a means for expressing the combinatorial compositional structure necessary for any system of linguistic representation as such.[45]

For Wittgenstein, there thus is a single privileged set of logical constants common to all possible systems of linguistic representation: the classical truth-functional connectives and quantifiers. Hence classical truth-functional and quantificational logic is the only conceivable possibility for expressing the "*logic* of the facts." By contrast, "[Wittgenstein's] absolutistic conception of language, in which the conventional element in the construction of a language is overlooked" is precisely what Carnap's own explanation of the logical constants and the resulting notion of logical or analytic truth

44 Goldfarb and Ricketts (1992, p. 68) present a minimalist reading of Carnap's logicism as simply the proposal for a "total language, which contains both logico-mathematical and synthetic sentences" adequate for representing the application of mathematics (§84). By contrast, it is crucial to my interpretation that the "logico-mathematical" or analytic sentences have a formally specifiable general character in virtue of which they are thereby entirely independent of all "matters of fact."

45 *Tractatus*, 5.4, 5.441, 5.47. For further discussion of these ideas, in connection with Carnap's conception of the logical constants sketched below, see Chapter 8 (this volume).

aims to avoid (§52). In particular, Carnap, in accordance with his principle of tolerance, wants to allow both underlying logics differing from classical logic and, in the case of classical mathematics itself, an expansion of the logical constants to include the identity sign, the numerals taken as primitive signs, and the full higher-order apparatus of classical analysis and set theory. For both of these reasons, therefore, Wittgenstein's minimalist, purely combinatorial conception of the logical constants is clearly inadequate. How, then, can Carnap continue to profess allegiance to the Wittgensteinian doctrine of tautology? How can he continue to maintain that the meanings of the logical constants – now explicitly *relativized* to the choice of one or another formal language or linguistic framework – are entirely independent of all "matters of fact"?

Section 50 of *Logical Syntax* contains Carnap's answer to these questions. For Carnap here presents, in general syntax, a characterization of the distinction between logical and descriptive expressions that is to hold for any possible formal language or linguistic framework and is intended formally to represent the idea that the logical expressions, relative to any given framework, are entirely independent of all extralinguistic factors:

> If a material interpretation is given for the language L, then one can divide the signs, expressions, and sentences of L into logical and descriptive, namely, into those with purely logico-mathematical meaning and those that signify something extra-logical, e.g., empirical objects or properties or the like. This classification is not only unsharp, but it is also non-formal, and thus not usable in syntax. If, however, we reflect that all interconnections of logico-mathematical concepts are independent of extra-linguistic determinations, e.g., empirical observations, and must be already completely fixed solely by the transformation rules of the language, we then find that the formally comprehensible distinguishing peculiarity of the logical signs and expressions is the circumstance that every sentence constructed from them alone is determinate.

The transformation rules of a language include, in general, both logical and physical rules, so not every sentence determined by the transformation rules is an analytic sentence. But the logico-mathematical expressions (in contradistinction to descriptive expressions like the electromagnetic-field functor) are such that *everything* about their use is already predetermined by the transformation rules (whereas some particular sentences containing the electromagnetic-field functor, for example, are not determined by the transformation rules even in the presence of Maxwell's equations). And it is in this precise and formal sense, therefore, that the logical expressions are independent of all extralinguistic factors or "matters of fact."

Carnap's formal characterization of the logical signs thereby transforms and replaces both the vague and intuitive conception of expressions that fail to "signify something extra-logical" and Wittgenstein's minimalist and "absolutistic" purely combinatorial conception. It allows us, in a precise and formal way, to harmonize the relativity to language encapsulated in Carnap's principle of tolerance with Wittgenstein's insight into the utter independence of logic from all "contingent facts of the world."[46] And, at the same time, it gives precise and formal expression to the fundamental idea that the truths of logic and mathematics are true solely in virtue of the meanings of the terms they contain. Analytic sentences, in contrast to synthetic sentences, contain only logical expressions essentially (§51), and so, their truth can be due only to the latter. But, in the case of logical expressions, everything about their use is already predetermined by the transformation rules and is in this sense purely linguistic. So here we have the best possible case of truth in virtue of meaning alone.[47] Finally, by transforming and replacing the intuitive, pretheoretical distinction between those expressions that signify empirical objects and properties and those that do not, Carnap's formal characterization makes it clear that he is not caught in the predicament depicted by Gödel discussed in §I (this chapter). Carnap does not take for granted a realm of empirical facts somehow intuitively given, but rather formally characterizes – relative to one or another formal language or linguistic framework – the very distinction between the formal and the factual itself.[48]

46 Note here that Carnap thereby articulates a sense in which the truths of logic and mathematics are "empty of content" without invoking protocol sentences and thus a notion of *empirical* content. That sentences essentially containing descriptive expressions should be testable via the deduction of protocol sentences is then a separate requirement, which serves to differentiate legitimate (theoretical) descriptive concepts such as the electromagnetic-field functor from illegitimate descriptive concepts such as "entelechy" (§82, p. 319[247]). The requirement that P-rules, in particular, should be testable is thus in no way a trivial one.

47 In this sense, there is a notion of truth in virtue of meaning already in *Logical Syntax*, and Carnap's later use of this notion in the remarks cited in footnote 40, this chapter, applies equally here. More generally, Carnap's project in *Logical Syntax* is not to reject all talk about meaning, but rather to translate or reinterpret such talk in purely syntactical terms (as he does in §50 itself, and also, for example, in §§62 and 75). Primarily for the purpose of formally developing inductive logic, Carnap will later introduce "meaning postulates" to capture what he takes to be meaning relations involving *descriptive* expressions as well. But this extension of the notion of truth in virtue of meaning does not affect Carnap's philosophy of logic and mathematics (where, as I say in the text, we have the best possible case).

48 I thus find myself in substantial agreement with Ricketts' (1994, pp. 189–93) discussion of §50 of *Logical Syntax* and its implications for Gödel's objection. Once again, however, for Ricketts, what is of paramount importance is the problem in descriptive syntax of "finding" a

VI

From this point of view, Carnap's formal characterization of the distinction between logical and descriptive expressions in §50 of *Logical Syntax* bears considerable philosophical weight. Although Carnap has indeed given up the traditional logicist project of reducing classical mathematics to logic in some antecedently understood sense, he maintains, nonetheless, that classical mathematics consists only of analytic truths and is thus entirely independent of the facts of the actual world. Moreover, it is logicism in precisely this sense that then allows him to apply the principle of tolerance to the choice of logico-mathematical rules (whether classical or otherwise), which choice is now seen as concerning only the "linguistic form" of our total scientific system rather than its content. And, in particular, the choice of logico-mathematical rules (including the strong existential assumptions of classical arithmetic, analysis, and set theory) is now seen to have no ontological implications whatsoever.[49] Nor need Carnap accept Gödel's demand for a (nontrivial) consistency proof for the logico-mathematical rules in order to show that they do not lead to unintended empirical consequences. Since Carnap's own version of empiricism is simply the requirement that synthetic sentences essentially containing descriptive expressions should be testable via the deduction of further synthetic sentences (protocol

given formal language or linguistic framework in the speech dispositions of actual speakers, and, from this point of view, the notion of what speakers are committed to simply in virtue of the adoption of a framework – which includes both logico-mathematical rules (L-rules) and physical rules (P-rules) – is more important than the distinction, within a given framework, between logico-mathematical and physical rules (see footnotes 39 and 42, this chapter). Accordingly, Ricketts again takes the point of §50 to be entirely deflationary – it "displaces, more than analyzes, the notion of truth-in-virtue-of" (p. 191). On my reading, by contrast, the point of §50 is precisely to *explicate* the pretheoretical notion of truth in virtue of meaning (compare footnotes 40 and 47, this chapter). This is one notion that Carnap does not at all wish to deflate, on pain of undermining both his commitment to logicism and the principle of tolerance.

49 The last sentence of Carnap(1950a/1956, §5, p. 221) recasts the principle of tolerance so as to emphasize the duality between "linguistic forms" (external questions involving analytic sentences) and genuine "assertions" (internal questions). And in §3 (p. 215, n. 5), after referring to Paul Bernays's "Sur le platonisme dans les mathématiques" (a penetrating discussion of the foundations of logic and mathematics from a frankly ontological point of view), Carnap observes that "Quine does not acknowledge the distinction which I emphasize above [between internal and external questions], because according to his general conception there are no sharp boundary lines between logical and factual truth, between questions of meaning and questions of fact, between the acceptance of a language structure and the acceptance of an assertion formulated in the language." This passage strongly suggests, I believe, that Carnap himself sees a close connection between the analytic/synthetic distinction and the distinction between internal and external questions (see footnote 39, this chapter).

sentences), logico-mathematical rules themselves – which, by definition, contain no descriptive expressions – cannot possibly have empirical consequences (see footnote 11, this chapter). And security against inconsistency, according to the principle of tolerance, is simply one more pragmatic virtue among others.[50]

Yet Carnap's formal characterization of the distinction between logical and descriptive expressions also poses serious problems for his principle of tolerance. That principle bids us to view the dispute in the foundations of logic and mathematics between logicism, formalism, and intuitionism as a purely pragmatic question of which logico-mathematical rules we wish to adopt as the "linguistic form" of our total system of science. It appears, then, that in order properly to address this reformulation of the dispute, we should first step back from the decision itself so as to investigate impartially the formal consequences of each and every option:

> It is important to be aware of the conventional components in the construction of a language system. This view leads to an unprejudiced investigation of the various forms of new logical systems which differ more or less from the customary form (e.g., the intuitionistic logic constructed by Brouwer and Heyting, the systems of logic of modalities constructed by Lewis and others, the systems of plurivalued logic as constructed by Lukasiewicz and Tarski, etc.), and it encourages the construction of further new forms. The task is not to decide which of the different systems is "the right logic" but to examine their formal properties and the possibilities for their interpretation and application in science. It might be that a system deviating from the ordinary form will turn out to be useful as a basis for the language of science. (Carnap 1939, §12)

Hence, on a very natural understanding of the principle of tolerance, before we make any substantial decision about the logico-mathematical form of the language of science, we are to engage in a prior investigation, from a neutral and impartial vantage point, of the syntactic consequences of each and every "linguistic form" under consideration.[51]

50 Moreover, the distinction between logical and descriptive expressions gives Carnap a sense in which logico-mathematical rules are empty of content that does not rely on his empiricism (footnote 46, this chapter). Carnap thus is not vulnerable to a second objection leveled by Gödel, namely, that Carnap's notion of content arbitrarily begs the question in favor of empiricism (1995, pp. 354–5). By the same token, however, this further underscores the importance, for Carnap, of giving a general explication of the distinction that *formally characterizes* the pretheoretical idea of independence from all extralinguistic facts. (I am indebted to Goldfarb for emphasizing the importance of this second Gödelian objection here.)

51 Compare also the passage from §44 of *Logical Syntax* cited in footnote 25, this chapter, which also suggests an enterprise of "investigating consequences" logically prior to (pragmatic) "justification."

As we pointed out in §I (this chapter), however, a variant of Gödel's objection shows that Carnap's own metatheoretical standpoint cannot be neutral and impartial in this sense. Carnap's characterization of the distinction between logical and descriptive expressions requires, in the case of classical mathematics, that the consequence relation expressed in the transformation rules for the language of mathematical physics be non-recursively-enumerable (and indeed nonarithmetical). In giving a metatheoretical description of this language, we therefore need a meta-language even stronger than the language of classical mathematics itself (containing, in effect, classical mathematics plus a truth definition for classical mathematics). And we need this strong meta-language, not to prove the consistency of the classical linguistic framework in question, but simply to describe and define this framework in the first place so that questions about the consequences of adopting it (including the question of consistency) can then be systematically investigated. Even to begin to investigate this framework in logical syntax, then, we can in no way step back from the decision whether or not to adopt such a strong set of logico-mathematical rules. On the contrary, the only way in which we can describe this framework, in Carnap's terms, is to step up into an even stronger set of logico-mathematical rules, where the decision under consideration has itself already been made.

We also pointed out earlier that Carnap is perfectly clear about the technical situation, and he shows no qualms whatsoever about the use of such a strong meta-language (compare footnote 15, this chapter). Indeed, it might now seem, as Carnap explicitly states in §45, that the principle of tolerance should then apply, in turn, to the choice of meta-language as well, so that no conflict with this principle could possibly arise here.[52] But, from our present point of view, the situation is not so simple. Consider, for example, the choice between classical logico-mathematical rules for the total language of science and the much weaker logico-mathematical rules endorsed by the intuitionist. To apply the principle of tolerance, we must view this choice as a purely pragmatic decision about "linguistic forms" having no ontological implications about "facts" or "objects" in the world. It is a matter of simply weighing one purely pragmatic virtue, ease of application, against a conflicting purely pragmatic value, safety against contradiction. Accordingly, we must view the logico-mathematical rules in question, in both linguistic

52 This idea, as we also have seen, is central to Goldfarb and Ricketts (1992). See especially (p. 69): "Carnap would surely disavow any pretense that there is one metalanguage that will always be acceptable to all parties in all controversies: there is no more a universal metalanguage than there is a universal object language."

frameworks, as sets of purely analytic sentences. Given Carnap's own explication of the distinctions between logical and descriptive terms, analytic and synthetic sentences, however, we must have already adopted the classical logico-mathematical rules in the meta-language. Thus, to understand the choice between classical and intuitionistic logico-mathematical rules in accordance with the principle of tolerance, we must have already built the former logico-mathematical rules into our background syntactic metaframework. We must have already biased the choice against the intuitionist in the very way in which we have set up the problem. The principle of tolerance, on Carnap's own understanding of it, then appears to undermine itself.[53]

In particular, the principle of tolerance by no means yields an initial situation of equal opportunity, where we are then free to adopt any of the positions in question in light of how they fare with respect to one or another set of purely pragmatic virtues. On the contrary, in the case of the philosophical debate in the foundations of mathematics that the principle was originally intended to dissolve, the very decision at issue has itself been already prejudged.[54] By contrast, the logicist side of Carnap's position appears to be completely self-consistent – and even, in a way, self-supporting. According to Carnap's logicism, we are urged, despite the possibility of contradiction, to adopt the full strength of the classical logico-mathematical rules. And we are told, in addition, that these rules are purely analytic truths, which thus function as mere "formal auxiliaries" having no ontological import for the "objects" and "facts" in the world. Then, by adopting the classical logico-mathematical rules in the meta-language as well, we can employ Carnap's formal characterization of the distinctions between logical and descriptive expressions, analytic and synthetic sentences, to cash out – and indeed to prove – these philosophical claims via translations into the

53 Goldfarb and Ricketts (1992, pp. 69–70) explicitly consider this situation, where they again see no problem for Carnap. What they fail to consider, from the present point of view, is how the principle of tolerance involves Carnap's (mathematically very strong) concept of analyticity, so that the pragmatically motivated intuitionist also then will be barred from embracing this principle as Carnap understands it. In my opinion, Goldfarb and Ricketts are here, once again, failing properly to appreciate the absolutely central role of Carnap's distinctions between logical and descriptive expressions, analytic and synthetic sentences, for the principle of tolerance itself (compare footnotes 42, 44, and 48, this chapter). Note that the pragmatically motivated intuitionist will eschew the use of a strong meta-language for the very same pragmatic reason that weighs against the choice of a strong object language, namely, the cautious desire to be as safe as possible against contradiction.

54 In a different context, Goldfarb and Ricketts (1992, p. 76) remark that "the method of logical syntax is meant to provide a level playing field for all contested views." If the present interpretation is correct, however, this is precisely what Carnap's understanding of the principle of tolerance does not do.

formal mode in logical syntax. If we are willing to adopt classical mathematics as the background logic of our metaframework, then we can prove, at least to ourselves, that this particular choice of logic is indeed analytic.[55]

Yet a Carnapian proponent of classical mathematics can also prove to himself – by precisely the argument sketched above – that his logicism stands in conflict with his tolerance. He can show that the mere idea that classical mathematics is analytic itself rules the intuitionist out of court. By contrast, the choice of a restricted meta-language equally acceptable to all parties to the dispute is much better suited to Carnap's profession of tolerance. In such a meta-language, we can still show that classical logico-mathematical rules (now described by a recursive proof relation rather than a non-recursively-enumerable consequence relation) are much stronger than the intuitionistic rules, so that, for example, the mean value theorem is easily provable in the former framework but not the latter. We thus can see, even in this restricted metaframework, that the classical rules are much more expedient for physical applications, whereas the intuitionistic rules provide far more safety against contradiction. Hence, in accordance with the spirit of the principle of tolerance, we then can view the choice between the two frameworks as a fundamentally pragmatic one. The only step we cannot take is to adopt Carnap's characteristic philosophical concept of analyticity so as to find translations in the formal mode of the philosophical claims constituting Carnap's logicism. We cannot set up a sharp contrast between *merely* pragmatic questions of "linguistic form" having no ontological import, on the one side, and genuine theoretical claims, on the other.[56] It is in this precise sense that the spirit of the principle of tolerance stands in conflict

55 This situation may be what Goldfarb and Ricketts (1992, p. 71) have in mind when they write that "[Carnap's] position is not circular so much as self-supporting at each level. If the mathematical part of a framework is analytic, then it's analytic; and so invoking mathematical truths at the level of the metalanguage is perfectly acceptable, since they flow from the adoption of the metalanguage."

56 We are now in a position to pinpoint precisely the mistake I made in my earlier treatment of these issues (Chapter 7, this volume, originally appearing in 1988). I there argued that Carnap's logicism needs to respect Wittgenstein's conception of the logical constants and that Carnap is thereby committed to a minimalist, purely combinatorial conception of logic in the meta-language. However, whereas Carnap does need to respect Wittgenstein's conception of the logical constants, the whole point of logical syntax, in this regard, is to generalize and relativize Wittgenstein's conception in the manner of §50. And this generalized conception does not lead to a minimalist version of logical syntax in the meta-language, but rather to an extremely strongly classical version. It is then the spirit of the principle of tolerance, not Carnap's logicism, that pushes us toward a minimal version of logical syntax, so that, in the end, Carnap's logicism stands in conflict with his tolerance.

with Carnap's logicism – and therefore, as we have seen, with the letter of that principle.[57]

VII

How damaging is this situation to Carnap's philosophical position? My own view is that it reveals a fundamental tension between his logicism and his tolerance that, in particular, renders his attempted dissolution of the philosophical debate in the foundations of logic and mathematics otiose. Yet Carnap himself never explicitly considers this problem. After accepting Tarski's theory of truth and adding the methods of formal semantics to logical syntax, Carnap (1942, §39) officially repudiates the characterization of the distinctions between logical and descriptive expressions, analytic and synthetic sentences, offered in §§50–2 of *Logical Syntax*.[58] On this basis, he also frankly acknowledges that, although he can still make the relevant distinctions for particular individual formal languages in "special semantics," he no longer has an overarching characterization in "general semantics." Accordingly (1942, §13), "[t]he problem of the nature of logical deduction and logical truth . . . can still not be regarded as completely solved."[59] Nevertheless, Carnap sees no *fundamental* problem here, and he remains hopeful, throughout his career, that the desired explication can and will be found.[60]

57 Thus, what I am calling the spirit of the principle of tolerance is better represented by the (much less philosophically loaded) pragmatic attitude of the "ordinary working scientist" discussed in §III, this chapter. It also fits Quine's opposing, naturalistic version of pragmatism. See, in particular, the conclusion of Quine (1963, pp. 405–6): "Now I am as impressed as Carnap with the vastness of what language contributes to science and to one's whole view of the world; and in particular I grant that one's hypothesis as to what there is, e.g., as to there being universals, is at bottom just as arbitrary and pragmatic a matter as one's adoption of a new brand of set theory or even a new system of bookkeeping. Carnap in turn recognizes that such decisions, however conventional, 'will nevertheless usually be influenced by theoretical knowledge.'[*] But what impresses me more than it does Carnap is how well this whole attitude is suited also to the theoretical hypotheses of natural science itself, and how little basis there is for a distinction." (The footnote is to §2 of "Empiricism, Semantics, and Ontology.")

58 Carnap's reasons for abandoning §§50–2 have nothing to do with the problems we have been discussing. The main reason appears to be that he now wants to recognize logical systems, such as first-order logic with identity, in which not all sentences containing only logical expressions are determinate. For an interesting discussion of the transition from syntax to semantics, containing a rather different suggestion as to what motivates Carnap to abandon §§50–2, see Ricketts (1996).

59 For the importance of a characterization in what Carnap is now calling "general semantics," see footnote 50, this chapter.

60 See, e.g., the quotation from Carnap (1963b) in footnote 43, this chapter.

Certainly, Carnap never sees any tension at all between the principle of tolerance and the analytic/synthetic distinction.[61]

As we have seen, Carnap aims to offer scientifically minded philosophers a systematic escape from their philosophical perplexities. We can systematically transform obscure and fruitless ontological disputes about the "reality" or "true nature" of some contested class of entities (such as numbers and other mathematical objects) into precise and fruitful disputes about the logico-linguistic form in which the total language of science is to be cast. We are thereby invited to recognize, in particular, that there is, after all, no genuine ontological import – no implications as to the "objects" and "facts" in the world – in the philosophical questions with which we have hitherto been struggling in vain. For, when we attain Carnapian philosophical self-consciousness, we see that we have actually been concerned with the much more fruitful, albeit purely pragmatic, question of language planning. In this way, Carnap's attempt to transform traditional philosophy into the new enterprise of language planning is intended to bring peace and progress to the discipline, much as his work on "the construction of an auxiliary language for international communication" was intended to contribute toward peace and progress for humankind in general.[62] It cannot be stressed too much, I believe, that Carnap himself was extraordinarily, and equally, serious about both of these ambitions.

In the end, what is perhaps most discouraging to Carnap's philosophical ambitions is the circumstance that his invitation to scientifically minded philosophers to transform their understanding of the discipline in this way has been almost universally ignored. A large number of philosophers, to be sure, have enthusiastically embraced the use of formal-logical methods,

61 Thus the problem referred to in footnote 58, this chapter, can be viewed as a narrowly technical rather than a truly fundamental one. (Moreover, from the point of view of *Logical Syntax*, this problem does not seem particularly important, since no one would seriously suggest logico-mathematical rules for the total language of science consisting solely of first-order logic with identity.) The closest Carnap comes to recognizing the problems with which we have been occupied is in his exchange with Beth in the Schilpp volume. Beth (1963, p. 479; see also pp. 499–502) uses ideas very close to our considerations to suggest that the need for a strong meta-language entails a "limitation regarding the Principle of Tolerance." In his reply, Carnap (1963b, pp. 929-30) explains that in a dispute between two parties touching also the question of the meta-language: "It may be the case that one of them can express in his own language certain convictions which he cannot translate into the common language; in this case he cannot communicate these convictions to the other man. For example, a classical mathematician is in this situation with respect to an intuitionist . . . ." Yet Carnap never takes up Beth's theme of a "limitation regarding the Principle of Tolerance."

62 See again (1963a, §11) for Carnap's parallel interest in both forms of language planning.

many of which were first pioneered by Carnap himself. Yet such philosophers, on the whole, have not simultaneously embraced Carnap's particular conception of the wider philosophical significance of these formal-logical methods. They have not come to conceive their enterprise as a purely pragmatic exercise in language planning having no theoretical or ontological implications whatsoever. Indeed, the three scientifically minded philosophers who worked most closely with Carnap during the formulation and elaboration of his *Logical Syntax* project – Gödel, Tarski, and Quine – all came explicitly to oppose Carnap's philosophical position. All three appeared to take considerations very close to those on which we have been focusing to constitute formidable, if not fatal, obstacles to Carnap's philosophical project. Gödel, as we have seen in §I, this chapter, took problems associated with the need for a strong meta-language in the light of his incompleteness results to pose a conclusive refutation of Carnap's philosophy of mathematics. Tarski opposed a sharp distinction between logical and descriptive expressions, and, on this basis, he publicly joined with Quine in rejecting the analytic/synthetic distinction.[63] And Quine, by far the most important philosopher among Carnap's students, appealed to the technical problems surrounding §§50–2 of *Logical Syntax* that we have explored here in carrying out a full-scale attack on all of the most fundamental notions of Carnap's philosophical framework, an attack that led to the widespread promulgation of a naturalistic form of pragmatism wherein Carnap's most cherished Fregean distinction – that between logical and psychological investigations – eventually fell by the wayside as well.[64] Carnap's invitation radically to transform the philosophical enterprise, an invitation deeply rooted, as we have seen, in a radically new conception of the debate in the philosophical foundations of logic and mathematics, could not have produced a more disappointing result.

63 For Carnap's relations with Tarski in connection with these issues, see Carnap (1963a, pp. 13, 30–1, 35–6, 60–7). See further Beth (1963, pp. 482–8), together with Carnap's reply (1963b, pp. 931–2).

64 For Quine's discussion of §§50–2 of *Logical Syntax*, see Quine (1963, §VII). For Carnap's view of the importance of a sharp distinction between logical and psychological investigations see, e.g., Carnap (1934a, p. 6; 1934c/1937, §72; 1934b, §I; 1935a, pp. 31–4; 1936; 1950b/1962, §§11–12).

BIBLIOGRAPHY

ASP (Archives for Scientific Philosophy). University of Pittsburgh Libraries. References are to file folder numbers. All rights reserved.

Ayer, Alfred Jules. 1936. *Language, Truth and Logic.* London: Gollancz.

Ayer, Alfred Jules (ed.). 1959. *Logical Positivism.* New York: Free Press.

Bauch, Bruno. 1923. *Wahrheit, Wert und Wirklichkeit.* Leipzig: Meiner.

Becker, Oskar. 1923. "Beiträge zur phänomenologischen Begründung der Geometrie und ihrer physikalischen Anwendungen." *Jahrbuch für Philosophie und phänomenologische Forschung* VI: 385–560.

Benacerraf, Paul, and Hilary Putnam (eds.). 1983. *Philosophy of Mathematics: Selected Readings.* 2nd ed. Cambridge, UK: Cambridge University Press.

Beth, Evert W. 1963. "Carnap's Views on the Advantages of Constructed Language Systems." In Schilpp (1963), pp. 469–502.

Carnap, Rudolf. 1922. *Der Raum. Ein Beitrag zur Wissenschaftslehre,* Kant-Studien Ergänzungsheft No. 56. Berlin: Reuther & Reichard.

1923. "Über die Aufgabe der Physik und die Anwendung des Grundsatzes der Einfachstheit." *Kant-Studien* 28: 90–107.

1924. "Dreidimensionalität des Raumes und Kausalität." *Annalen der Philosophie und philosophischen Kritik* 4: 105–30.

1925. "Über die Abhängigkeit der Eigenschaften des Raumes von denen der Zeit." *Kant-Studien* 30: 331–45.

1928a. *Der logische Aufbau der Welt.* Berlin: Weltkreis. 2nd ed. Hamburg: Meiner, 1961. Translated by R. George as *The Logical Structure of the World.* Berkeley and Los Angeles: University of California Press, 1967.

1928b. *Scheinprobleme in der Philosophie.* Berlin: Weltkreis. Translated by R. George as *Pseudoproblems in Philosophy.* Berkeley and Los Angeles: University of California Press, 1967.

1929. *Abriss der Logistik, mit besonderer Berücksichtigung der Relationstheorie und ihrer Anwendungen.* Wien: Springer.

1930a. "Die alte und die neue Logik." *Erkenntnis* 1: 12–26. Translated by I. Levi as "The Old and the New Logic." In Ayer (1959), pp. 133–46.

1930b. "Die Mathematik als Zweig der Logik." *Blätter für deutsche Philosophie* 4: 298–310.

1931a. "Die logizistische Grundlegung der Mathematik." *Erkenntnis* 2: 91–105. Translated by E. Putnam and G. J. Massey as "The Logicist Foundation of Mathematics." In Benacerraf and Putnam (1983), pp. 41–52.

1931b. "Diskussion zur Grundlegung der Mathematik." *Erkenntnis* 2: 141–4.

1932a. "Überwindung der Metaphysik durch logische Analyse der Sprache." *Erkenntnis* 2: 219–41. Translated by A. Pap as "The Elimination of Metaphysics Through Logical Analysis of Language." In Ayer (1959), pp. 60–81.

1932b. "Die physikalische Sprache als Universalsprache der Wissenschaft." *Erkenntnis* 2: 432–65. Translated by M. Black as *The Unity of Science.* London: Kegan Paul, 1934.

1932c. "Psychologie in physikalischer Sprache." *Erkenntnis* 3: 107–42. Translated by G. Schick as "Psychology in Physical Language." In Ayer (1959), pp. 165–98.

1932d. "Über Protokollsätze." *Erkenntnis* 3: 215–28. Translated by R. Creath and R. Nollan as "On Protocol Sentences." *Noûs* 21 (1987): 457–70.

1934a. "On the Character of Philosophical Problems." *Philosophy of Science* 1: 2–19.

1934b. *Die Aufgabe der Wissenschaftslogik.* Wien: Gerold. Translated by H. Kaal as "The Task of the Logic of Science." In B. McGuinness (ed.), *Unified Science.* Dordrecht: Reidel, 1987, pp. 46–66.

1934c. *Logische Syntax der Sprache.* Wien: Springer. Translated by A. Smeaton as *The Logical Syntax of Language.* London: Kegan Paul, 1937.

1935a. *Philosophy and Logical Syntax.* London: Kegan Paul.

1935b. "Formalwissenschaft und Realwissenschaft." *Erkenntnis* 5: 30–7. Translated by H. Feigl and M. Brodbeck as "Formal and Factual Science." In H. Feigl and M. Brodbeck (eds.), *Readings in the Philosophy of Science.* New York: Appleton-Century-Crofts, 1953, pp. 123–8.

1936. "Von der Erkenntnistheorie zur Wissenschaftslogik." *Actes du Congrès International de Philosophie Scientifique,* vol. 1. Paris: Hermann, 36–41.

1939. *Foundations of Logic and Mathematics. International Encyclopedia of Unified Science,* vol. I, no. 3. Chicago: University of Chicago Press.

1942. *Introduction to Semantics.* Cambridge, Mass.: Harvard University Press.

1950a. "Empiricism, Semantics, and Ontology." *Revue Internationale de Philosophie* 11: 20–40. Reprinted in *Meaning and Necessity.* 2nd ed. Chicago and London: University of Chicago Press, 1956, pp. 205–29.

1950b. *Logical Foundations of Probability.* 2nd ed. 1962. Chicago: University of Chicago Press.

1954. *Einführung in die symbolische Logik, mit besonderer Berücksichtigung ihrer Anwendungen.* Wien: Springer. Translated by W. Meyer and J. Wilkinson as *Introduction to Symbolic Logic and Its Applications.* New York: Dover, 1958.

1963a. "Intellectual Autobiography." In Schilpp (1963), pp. 3–84.

1963b. "Replies and Systematic Expositions." In Schilpp (1963), pp. 859–1013.

1966. *Philosophical Foundations of Physics: An Introduction to the Philosophy of Science,* ed. M. Gardner. New York: Basic Books. Reprinted as *An Introduction to the Philosophy of Science,* 1974.

Cassirer, Ernst. 1907. "Kant und die moderne Mathematik." *Kant-Studien* 12: 1–49.

1910. *Substanzbegriff und Funktionsbegriff.* Berlin: Bruno Cassirer. Translated by W. Swabey and M. Swabey as *Substance and Function.* Chicago: Open Court, 1923.

1913. "Erkenntnistheorie nebst den Grenzfragen der Logik." *Jahrbücher der Philosophie* 1: 1–59.

1921. *Zur Einsteinschen Relativitätstheorie.* Berlin: Bruno Cassirer. Translated by W. Swabey and M. Swabey as *Einstein's Theory of Relativity.* Chicago: Open Court, 1923.

1929. *Philosophie der symbolischen Formen: III. Phänomenologie der Erkenntnis.* Berlin: Bruno Cassirer. Translated by R. Manheim as *The Philosophy of Symbolic Forms, Volume Three: The Phenomenology of Knowledge.* New Haven: Yale University Press, 1957.

1932. *Die Philosophie der Aufklärung.* Tübingen: Mohr. Translated by F. Koelln and J. Pettegrove as *The Philosophy of the Enlightenment.* Princeton, N.J.: Princeton University Press, 1951.

Coffa, J. Alberto. 1987. "Carnap, Tarski and the Search for Truth." *Noûs* 21: 547–72.

1991. *The Semantic Tradition from Kant to Carnap: To the Vienna Station.* Cambridge, UK: Cambridge University Press.

Creath, Richard. 1982. "Was Carnap a Complete Verificationist in the Aufbau?" *PSA 1982,* vol. 1, pp. 384–93.

1990. *Dear Carnap, Dear Van: The Quine-Carnap Correspondence and Related Work*. Berkeley and Los Angeles: University of California Press.

Demopoulos, William, and Michael Friedman. 1985. "Bertrand Russell's *Analysis of Matter*: Its Historical Context and Contemporary Interest." *Philosophy of Science* 52: 621–39.

Einstein, Albert. 1916. "Die Grundlage der allgemeinen Relativitätstheorie." *Annalen der Physik* 49: 769–822. Translated by W. Perrett and G. Jeffrey as "The Foundation of the General Theory of Relativity." In H. A. Lorentz (ed.), *The Principle of Relativity*. London: Methuen, 1923, pp. 111–64.

1917. *Über die spezielle und die allgemeine Relativitätstheorie, gemeinverständlich*. Braunschweig: Vieweg. Translated by R. Lawson from the 5th ed. as *Relativity, the Special and the General Theory: A Popular Exposition*. London: Methuen, 1920.

1921. "Geometrie und Erfahrung." *Preussische Akademie der Wissenschaft. Physikalisch-mathematische Klasse. Sitzungsberichte*, pp. 123–30. Expanded and translated as "Geometry and Experience." In G. Jeffrey and W. Perrett (eds.), *Sidelights on Relativity*. London: Methuen, 1923, pp. 27–55.

Feferman, S., J. Dawson, W. Goldfarb, C. Parsons, and R. Solovay (eds.). 1995. *Kurt Gödel: Collected Works*, vol. III. Oxford: Oxford University Press.

Frank, Philipp. 1949. *Modern Science and Its Philosophy*. Cambridge, Mass.: Harvard University Press.

Frege, Gottlob. 1879. *Begriffsschrift, eine der arithmetischen nachgebildete Formelsprache des reinen Denkens*. Halle. Translated by S. Bauer-Mengelberg as "*Begriffsschrift*, a formula language, modeled upon that of arithmetic, for pure thought." In van Heijenoort (1967), pp. 5–82.

Friedman, Michael. 1983. *Foundations of Space-Time Theories: Relativistic Physics and Philosophy of Science*. Princeton, N.J.: Princeton University Press.

1992a. *Kant and the Exact Sciences*. Cambridge, Mass.: Harvard University Press.

1992b. "Causal Laws and the Foundations of Natural Science." In P. Guyer (ed.), *The Cambridge Companion to Kant*. Cambridge, UK: Cambridge University Press, pp. 161–99.

1992c. "Philosophy and the Exact Sciences: Logical Positivism as a Case Study." In J. Earman (ed.), *Inference, Explanation, and Other Frustrations*. Berkeley and Los Angeles: University of California Press, pp. 84–98.

1996. "Overcoming Metaphysics: Carnap and Heidegger." In Giere and Richardson (1996), pp. 45–79.

1997. "Helmholtz's *Zeichentheorie* and Schlick's *Allgemeine Erkenntnislehre*: Early Logical Empiricism and Its Nineteenth Century Background." *Philosophical Topics* 25: 19–50.

1999. "Geometry, Construction, and Intuition in Kant and His Successors." In G. Scher and R. Tieszen (eds.), *Between Logic and Intuition: Essays in Honor of Charles Parsons.* Cambridge, UK: Cambridge University Press.

Giere, Ronald. 1988. *Explaining Science: A Cognitive Approach.* Chicago: University of Chicago Press.

Giere, Ronald, and Alan Richardson (eds.). 1996. *Origins of Logical Empiricism.* Minneapolis: University of Minnesota Press.

Gödel, Kurt. 1931. "Über formal unentscheidbare Sätze der Principia mathematica und verwandter Systeme I." *Monatshefte für Mathematik und Physik* 38: 173–98. Translated as "On Formally Undecidable Propositions of *Principia Mathematica* and Related Systems I." In van Heijenoort (1967), pp. 596–616.

1995. "Is Mathematics Syntax of Language? [Versions III, V]." In Feferman, et al. (1995), pp. 334–62.

Goldfarb, Warren. 1979. "Logic in the Twenties: The Nature of the Quantifier." *Journal of Symbolic Logic* 44: 351–68.

1995. "Introductory Note to [Gödel 1995]." In Feferman, et al. (1995), pp. 324–34.

Goldfarb, Warren, and Thomas Ricketts. 1992. "Carnap and the Philosophy of Mathematics." In D. Bell and W. Vossenkuhl (eds.), *Science and Subjectivity.* Berlin: Akademie, pp. 61–78.

Goodman, Nelson. 1951. *The Structure of Appearance.* Cambridge, Mass.: Harvard University Press.

1963. "The Significance of *Der logische Aufbau der Welt.*" In P. Schilpp (1963), pp. 545–58.

Haack, Susan. 1977. "Carnap's *Aubau*: Some Kantian Reflexions." *Ratio* 19: 170–6.

Haller, Rudolf. 1985. "Der erste Wiener Kreis." *Erkenntnis* 22: 341–58. Translated by T. Uebel as "The First Vienna Circle." In Uebel (1991), pp. 95–108.

1993. *Neopositivismus: Eine historische Einführung in die Philosophie des Wiener Kreises.* Darmstadt: Wissenschaftliche Buchgesellschaft.

Hertz, Paul, and Moritz Schlick (eds.). 1921. *Hermann v. Helmholtz: Schriften zur Erkenntnistheorie.* Berlin: Springer. Translated by M. Lowe, ed. R. Cohen and Y. Elkana, as *Hermann von Helmholtz: Epistemological Writings.* Dordrecht: Reidel, 1977.

Hilbert, David. 1899. *Grundlagen der Geometrie.* Leipzig: Teubner. Translated by L. Unger from the 10th ed. as *Foundations of Geometry.* La Salle: Open Court, 1971.

Howard, Don. 1984. "Realism and Conventionalism in Einstein's Philosophy of Science: The Einstein-Schlick Correspondence." *Philosophia Naturalis* 21: 619–29.

Husserl, Edmund. 1913. *Ideen zu einer reinen Phänomenologie und phänomenologischen Philosophie. Erstes Buch: Allgemeine Einführung in die reine Phänomenologie.* Halle: Max Niemeyer. Translated by W. Boyce Gibson as *Ideas: General Introduction to Pure Phenomenology.* London: Allen & Unwin, 1931.

Kraft, Viktor. 1950. *Der Wiener Kreis.* Wien: Springer. Translated by Arthur Pap as *The Vienna Circle.* New York: Philosophical Library, 1953.

Kuhn, Thomas. 1962. *The Structure of Scientific Revolutions.* Chicago: University of Chicago Press.

Lewis, Clarence Irving. 1929. *Mind and the World-Order.* New York: Scribner.

Lotze, Hermann. 1874. *Logik.* Leipzig: Hirzel. Translated by B. Bosanquet as *Logic.* Oxford: Oxford University Press, 1884.

Mach, Ernst. 1886. *Beiträge zur Analyse der Empfindung.* Jena: Fischer. Translated by C. M. Williams. Revised and supplemented from the 5th ed. by S. Waterlow as *The Analysis of Sensations.* New York: Dover, 1959.

Menger, Karl. 1930. "Der Intuitionismus." *Blätter für deutsche Philosophie* 4: 311–25. Translated as "On Intuitionism." In *Selected Papers in Logic and Foundations, Didactics, and Economics.* Dordrecht: Reidel, 1979, pp. 46–58.

1994. *Reminiscences of the Vienna Circle and the Mathematical Colloquium,* ed. L. Golland, B. McGuinness, and A. Sklar. Dordrecht: Kluwer.

Moulines, C. Ulises. 1985. "Hintergründe der Erkenntnistheorie des frühen Carnap." *Grazer Philosophische Studien* 23: 1–18.

Mulder, Henk L., and Barbara F. B. van de Velde-Schlick (eds.). 1978–9. *Mortiz Schlick: Philosophical Papers.* 2 vols. Dordrecht: Reidel.

Natorp, Paul. 1910. *Die logischen Grundlagen der exakten Wissenschaften.* Leipzig: Teubner.

Oberdan, Thomas. 1993. *Protocols, Truth and Convention.* Amsterdam: Rodopoi.

Poincaré, Henri. 1902. *La Science et l'Hypothèse.* Paris: Flammarion. Translated by W. J. Greenstreet as *Science and Hypothesis.* London: Scott, 1905. Also translated by G. B. Halsted. In *The Foundations of Science.* Lancaster: Science Press, 1946.

Putnam, Hilary. 1975. "Language and Philosophy." In *Philosophical Papers,* vol. 2. Cambridge: Cambridge University Press, pp. 1–32.

Quine, Willard van Orman. 1951. "Two Dogmas of Empiricism." *Philosophical Review* 60: 20–43. Reprinted in *From a Logical Point of View*. New York: Harper, 1963, pp. 20–46.

1963. "Carnap and Logical Truth." In Schilpp (1963), pp. 385–406.

1969. "Epistemology Naturalized." In *Ontological Relativity and Other Essays*. New York: Columbia University Press, pp. 69–90.

Reichenbach, Hans. 1920. *Relativitätstheorie und Erkenntnis Apriori*. Berlin: Springer. Translated by M. Reichenbach as *The Theory of Relativity and A Priori Knowledge*. Los Angeles: University of California Press, 1965.

1922. "Der gegenwärtige Stand der Relativitätsdiskussion." *Logos* 10: 316–78. Translated by M. Reichenbach as "The Present State of the Discussion on Relativity." In M. Reichenbach and R. Cohen (eds.), *Hans Reichenbach: Selected Writings, 1909–1953*, vol. 2. Dordrecht: Reidel, 1978, pp. 3–47.

1924. *Axiomatik der relativistischen Raum-Zeit-Lehre*. Braunschweig: Vieweg. Translated by M. Reichenbach as *Axiomatization of the Theory of Relativity*. Los Angeles: University of California Press, 1969.

1928. *Philosophie der Raum-Zeit-Lehre*. Berlin: de Gruyter. Translated by M. Reichenbach and J. Freund as *The Philosophy of Space and Time*. New York: Dover, 1958.

Richardson, Alan. 1990. "How Not to Russell Carnap's Aufbau." *PSA 1990*, vol. 1, pp. 3–14.

1992. "Logical Idealism and Carnap's Construction of the World." *Synthese* 93: 59–92.

1996. "From Epistemology to the Logic of Science: Carnap's Philosophy of Empirical Knowledge in the 1930s." In Giere and Richardson (1996), pp. 309–32.

1998. *Carnap's Construction of the World: The Aufbau and the Emergence of Logical Empiricism*. Cambridge, UK: Cambridge University Press.

Rickert, Heinrich. 1892. *Der Gegenstand der Erkenntnis*. 3rd ed. 1915, 5th ed. 1921. Tübingen: Mohr.

1902. *Die Grenzen der naturwissenschaftlichen Begriffsbildung*. 5th ed. 1929. Tübingen: Mohr. Translated (partially) by G. Oakes as *The Limits of Concept Formation in Natural Science*. Cambridge, UK: Cambridge University Press, 1986.

Ricketts, Thomas. 1982. "Rationality, Translation, and Epistemology Naturalized." *Journal of Philosophy* 79: 117–37.

1985. "Frege, the *Tractatus*, and the Logocentric Predicament." *Noûs* 19: 3–15.

1994. "Carnap's Principle of Tolerance, Empiricism, and Conventionalism." In P. Clark and B. Hale (eds.), *Reading Putnam.* Oxford: Blackwell, pp. 176–200.

1996. "Carnap: From Logical Syntax to Semantics." In Giere and Richardson (1996), pp. 231–50.

Riemann, Bernhard. 1919. *Über die Hypothesen, welche der Geometrie zugrunde liegen. Neu herausgegeben und erläutert von H. Weyl.* Berlin: Springer.

Rorty, Richard. 1979. *Philosophy and the Mirror of Nature.* Princeton, N. J.: Princeton University Press.

Russell, Bertrand. 1903. *The Principles of Mathematics.* London: Allen & Unwin.

1914. *Our Knowledge of the External World as a Field for Scientific Method in Philosophy.* London: Allen & Unwin.

1919. *Introduction to Mathematical Philosophy.* London: Allen & Unwin.

1927. *The Analysis of Matter.* London: Allen & Unwin.

Ryckman, Thomas A. 1992. "'P(oint) C(oincidence) Thinking': The Ironical Attachment of Logical Empiricism to General Relativity (and Some Lingering Consequences)." *Studies in History and Philosophy of Science* 23: 471–93.

Sauer, Werner. 1985. "Carnaps 'Aufbau' in Kantianischer Sicht." *Grazer Philosophische Studien* 23: 19–35.

1987. "Carnaps Konstitutionstheorie und das Program der Einheitswissenschaft des Wiener Kreises." *Conceptus* 21: 233–45.

1989. "On the Kantian Background of Neopositivism." *Topoi* 8: 111–19.

Schilpp, Paul Arthur (ed.). 1963. *The Philosophy of Rudolf Carnap.* La Salle: Open Court.

Schlick, Moritz. 1915. "Die philosophische Bedeutung des Relativitätsprinzips." *Zeitschrift für Philosophie und philosophische Kritik* 159: 129–75. Translated by P. Heath as "The Philosophical Significance of the Principle of Relativity." In Mulder and van de Velde-Schlick (1978–9), vol. 1, pp. 153–89.

1917. *Raum und Zeit in der gegenwärtigen Physik.* Berlin: Springer. 3rd ed. 1920, 4th ed. 1922. Translated by H. Brose from the 3rd ed. as *Space and Time in Contemporary Physics.* Oxford: Oxford University Press, 1920. Expanded to include changes in the 4th ed. by P. Heath. In Mulder and van de Velde-Schlick (1978–9), vol. 1, pp. 207–69.

1918. *Allgemeine Erkenntnislehre.* Berlin: Naturwissenschaftliche Monographien und Lehrbücher. 2nd ed. Berlin: Springer, 1925. Translated by A. Blumberg as *General Theory of Knowledge.* La Salle: Open Court, 1985.

1921. "Kritizistische oder empiristische Deutung der neuen Physik?" *Kant-Studien* 26: 96–111. Translated by P. Heath as "Critical or Empiricist Interpretation of Modern Physics?" In Mulder and van de Velde-Schlick (1978–9), vol. 1, pp. 322–34.

1926. "Erleben, Erkennen, Metaphysik." *Kant-Studien* 31: 146–58. Translated by P. Heath as "Experience, Cognition and Metaphysics." In Mulder and van de Velde-Schlick (1978–9), vol. 2, pp. 99–111.

1932a. "Positivismus und Realismus." *Erkenntnis* 3: 1–31. Translated by P. Heath as "Positivism and Realism." In Mulder and van de Velde-Schlick (1978–9), vol. 2, pp. 259–84.

1932b. "Form and Content." Three Lectures: London. In Mulder and van de Velde-Schlick (1978–9), vol. 2, pp. 285–369.

1934. "Über das Fundament der Erkenntnis." *Erkenntnis* 4: 79–99. Translated by P. Heath as "On the Foundation of Knowledge." In Mulder and van de Velde-Schlick (1978–9), vol. 2, pp. 370–87.

Sluga, Hans. 1980. *Gottlob Frege.* Boston, London, and Henley: Routledge.

Tarski, Alfred. 1936. "Der Wahrheitsbegriff in den formalisierten Sprachen." *Studia Philosophica* 1: 261–405. Translated as "The Concept of Truth in Formalized Languages." In J. H. Woodger (ed.), *Logic, Sematics, Metamathematics: Papers from 1923 to 1938 by Alfred Tarski.* Oxford: Clarendon, 1956, pp. 152–278.

Uebel, Thomas (ed.). 1991. *Rediscovering the Forgotten Vienna Circle.* Dordrecht: Reidel.

Uebel, Thomas. 1992. *Overcoming Logical Positivism from Within: The Emergence of Neurath's Naturalism in the Vienna Circle's Protocol Sentence Debate.* Amsterdam: Rodopoi.

1996. "The Enlightenment Ambition of Epistemic Utopianism: Otto Neurath's Theory of Science in Historical Perspective." In Giere and Richardson (1996), pp. 91–112.

van Heijenoort, Jean (ed.). 1967. *From Frege to Gödel: A Source Book in Mathematical Logic, 1879–1931.* Cambridge, Mass.: Harvard University Press.

Waismann, Friedrich. 1967. *Wittgenstein und der Wiener Kreis.* ed. B. McGuinness. Frankfurt: Suhrkamp. Translated by B. McGuiness as *Wittgenstein and the Vienna Circle.* London: Blackwell, 1979.

Weyl, Hermann. 1918. *Raum-Zeit-Materie.* Berlin: Springer. 4th ed. 1921, 5th ed. 1922. Translated by H. Brose from the 4th ed. as *Space-Time-Matter.* London: Methuen, 1922.

1922. "Die Einzigartigkeit der Pythagoreischen Maßbestimmung." *Mathematische Zeitschrift* 12: 114–46.

1927. *Philosophie der Mathematik und Naturwissenschaft.* München: Oldenbourg. Revised and augmented English edition based on a translation by O. Helmer, *Philosophy of Mathematics and Natural Science.* Princeton: Princeton University Press, 1949.

Whitehead, Alfred North, and Bertrand Russell. 1910–13. *Principia Mathematica.* 3 vols. Cambridge, UK: Cambridge University Press.

Wittgenstein, Ludwig. 1922. *Tractatus Logico-Philosophicus,* with translation by D. Pears and B. McGuinness. London: Routledge.

INDEX

Printed in the United States
By Bookmasters